Regulating the Liabilities of Agricultural Biotechnology

This book is dedicated to:

Lanette, Connor, Reid and Kaelan Smyth,

Wanda, Alexander and Andrew Phillips,

May T. Yeung

and

Lorraine M. Khachatourians

Regulating the Liabilities of Agricultural Biotechnology

Stuart Smyth,[1]
Peter W.B. Phillips,[2]
William A. Kerr[2]
and
George G. Khachatourians[3]

[1]Department of Inter-disciplinary Studies,
[2]Department of Agricultural Economics
and
[3]Department of Applied Microbiology and Food Science
University of Saskatchewan
Canada

CABI Publishing

CABI Publishing is a division of CAB International

CABI Publishing
CAB International
Wallingford
Oxfordshire OX10 8DE
UK

CABI Publishing
875 Massachusetts Avenue
7th Floor
Cambridge, MA 02139
USA

Tel: +44 (0)1491 832111
Fax: +44 (0)1491 833508
E-mail: cabi@cabi.org
Web site: www.cabi-publishing.org

Tel: +1 617 395 4056
Fax: +1 617 354 6875
E-mail: cabi-nao@cabi.org

A catalogue record for this book is available from the British Library, London, UK.

Library of Congress Cataloging-in-Publication Data

Regulating the liabilities of agricultural biotechnology / Stuart Smyth ... [et al.]
    p. cm.
  Includes bibliographical references (p.   ).
  ISBN 0-85199-815-1 (alk. paper)
  1. Agricultural laws and legislation. 2. Agricultural biotechnology--Law and legislation. 3. Products liability--Food. 4. Agricultural biotechnology. I. Smyth, Stuart. II. Title.

K3870.R44 2004
343'.076--dc22

2004000335

ISBN 0 85199 815 1

Printed and bound in the UK by Biddles Ltd, King's Lynn, from copy supplied by the authors

Contents

List of Tables and Figures

List of Acronyms

AAFC	Agriculture and Agri-food Canada
ADM	Archer Daniels Midland
AIA	Advanced Informed Agreement
ANZ	Australia and New Zealand
APHIS	Animal and Plant Health Inspection Service
ASA	American Soybean Association
BIO	Biotechnology Industry Organization
B. napus	*Brassica napus*
B. rapa	*Brassica rapa*
BSE	Bovine Spongiform Encephalopathy
BSP	Biosafety Protocol
Bt	*Bacillus thuringiensis*
CAP	Common Agricultural Policy
CBD	Convention on Biological Diversity
CBI	Council for Biotechnology Information
CCC	Canola Council of Canada
CEC	Commission for Environmental Co-operation
CEO	Chief Executive Officer
CFIA	Canadian Food Inspection Agency
CGC	Canadian Grain Commission
CMS	Cytoplasmically Male-sterile
Codex	Codex Alimentarius Commission
COFFS	Canadian On-farm Food Safety
CSGA	Canadian Seed Growers Association
CSP	Cross-species Pollination
CWB	Canadian Wheat Board
DNA	Deoxyribonucleic Acid
DSM	Dispute Settlement Mechanism

DSM	Dispute Settlement Mechanism
E. coli	*Escherichia coli*
ELISA	Enzyme-Linked Immunosorbent Assay
ELSI	Ethics, Law, Society and Industry
EPA	Environmental Protection Agency
EU	European Union
FAO	Food and Agriculture Organisation
FDA	Food and Drug Administration
FIS	International Seed Trade Federation
f.o.b.	Freight on Board
GATT	General Agreement on Tariffs and Trade
GBP	Great Britain Pound
GDP	Gross Domestic Product
GFP	Green Fluorescent Protein
GLP	Good Laboratory Practices
GM	Genetically Modified
GMO	Genetically Modified Organism
GMP	Good Manufacturing Practices
GURTs	Genetic Use Restriction Technologies
HACCP	Hazard Analysis Critical Control Points
HEAR	High Erucic Acid Rapeseed
HT	Herbicide Tolerant
IAA	Indole Acetic Acid
ICCP	Intergovernmental Committee for the Cartagena Protocol on Biosafety
ICT	Information, Communications and Telecommunications
IFIC	International Food Information Council
IgE	Immunoglobulin E
IPE	International Political Economy
IPP	Identity Preserved Production

IPPM	Identity Preserved Production and Marketing
IPR	Intellectual Property Right
ISO	International Standards Organisation
ISP	Interspecific Pollination
ISPM	International Standards for Phytosanitary Measures
JRI	James Richardson International
LMO	Living Modified Organism
MMT	Methylcyclopentadienyl Manganese Tricarbonyl
NAFTA	North American Free Trade Agreement
MOU	Memorandum of Understanding
MRA	Members of Regional Association
NGO	Non-governmental Organisation
NIH	National Institutes of Health
OECD	Organisation for Economic Co-operation and Development
OIE	Office International des Epizooties
PGS	Plant Genetics Systems
PIAR	Pollen Induced Allergic Reactions
PMPs	Plant Made Pharmaceuticals
PNTs	Plants with Novel Traits
RAF	Risk Analysis Framework
R&D	Research and Development
rBST	Recombinant Bovine Somatotrophin
rDNA	Recombinant Deoxyribonucleic Acid
SOD	Saskatchewan Organic Directorate
SPS	Sanitary and Phytosanitary Agreement
SQC	Scottish Quality Cereals
SRA	Scientifically-based Risk Assessment
SWP	Saskatchewan Wheat Pool
TBT	Technical Barriers to Trade
TEP	Transatlantic Economic Partnership

TUA	Technology Use Agreement
UGG	United Grain Growers
UK	United Kingdom
UN	United Nations
UPOV	International Union for the Protection of New Varieties of Plants
US	United States
USDA	United States Department of Agriculture
WCCRRC	Western Canadian Canola Rapeseed Recommendation Committee
WHO	World Health Organisation
WTO	World Trade Organisation

Preface

The relationship between the legal community and agriculture has become considerably more complex over the past decade. The wide-ranging expansion of the previously limited intellectual property rights in living organisms has resulted in plants that have not only commercial market value but also proprietary knowledge value. To a considerable extent, these new plants are the result of the application of biotechnology. Commercial production of plants resulting from intellectual property has created a range of unique challenges for the agriculture industry. A number of these challenges have resulted in litigation, some of which are presently in front of the courts while others have already been settled.

When a risk is realised it becomes a liability. This has occurred on several occasions in agricultural biotechnology due to the co-mingling of transgenic crops with conventional or organic production. Additionally, some liabilities have resulted from regulatory violations committed by firms involved in the production of transgenic crop varieties. The management of liabilities in the commercial production of agricultural products has created a need to explain some of the challenges and to offer suggestions on how to best resolve the resulting liabilities. Hence, we have combined our research efforts to produce this book.

In the past, the stakeholders in genetic-based technological improvements have been government regulatory bodies and seed development firms or public research institutes. It is the interaction between the regulator and the developer of the new variety that has resulted in the commercialization of new crop varieties over the past several decades. The introduction of transgenic crops has dramatically changed the interaction between the traditional stakeholders and added some new interested parties.

In the early 1990s, the initial commercialization of transgenic crops provided the impetus for the inclusion of some new stakeholders. Within three years of the initial commercialization, concerned groups in civil society organized to oppose commercial production and the approval of new transgenic crops. Recently, the courts have become involved as a result of stakeholders seeking legal recourse to some issues involving transgenic crops. The need to include these two recent stakeholders in the assessment of transgenic crops has presented some interesting research questions.

Many of the opportunities to undertake research on topics relating to biotechnology are based in North American agriculture. The major reason for this geographic insularity is the moratorium on transgenic crops that has been in place in the European Union since 1998. The lack of any substantial commercial production in Europe has prevented the industry from maturing and its absence has resulted in most agricultural biotechnology applications being restricted to basic research and development.

As the maturing of the agricultural biotechnology industry unfolded in North America it was naturally accompanied by a number of questions pertaining to liability. Given the high level of transgenic crop research and adoption in Canada, and Western Canada in particular, there is a great deal of social research taking place. The result of this research has facilitated the coming into being of one of the world's leading social science research clusters examining the applications of agricultural biotechnology.

One result of this research cluster was that it provided the incentives to attract the four authors to the University of Saskatchewan. Collectively, the four authors of this book have been involved in researching agricultural biotechnology virtually since its inception. William Kerr has been researching and writing about agricultural biotechnology since 1989. Peter Phillips began his research efforts in this area in 1997 with the acceptance of a dedicated research chair at the University of Saskatchewan. Stuart Smyth has been extensively involved in this field of research since 1999 when he decided to return to university to undertake a Ph.D. in the area of socio-economic aspects of agricultural biotechnology. George Khachatourians has been involved with biotechnology since he sat on the Government of Canada's National Biotechnology Task Force in 1980-81.

Of course the production of a book required many hands and the authors would like to thank Marylou Langridge, Josefin Kihlberg and Julie Parchewski for their assistance in the preparation of the manuscript and Laura Loppacher for her work in compiling the index.

Stuart Smyth
Peter W.B. Phillips
William A. Kerr
George G. Khachatourians
Saskatoon, Canada, March 2004

Part I:

Introduction

Chapter One:

Liability and Transformative Technology

Introduction

Transformative technological change appears to both be spawned by and create chaos. Chaos is a word that brings forth strong emotions—fear, disgust, apprehension and, hopefully, for a very few, elation. Change is often confused with chaos. Change often brings forth the same emotions as chaos. The more rapid the rate of change, the more likely it is to look like chaos. The pace of change that characterises the convergence of new technologies that underlie the process that has been dubbed globalisation is very rapid. The spread of computing power to every corner of the developed world, including its enthusiastic uptake by children, has taken place in little more than a decade and a half. The electronic revolution in information technology embodied in the Internet has become a mass technology in half that time. The potential for these technologies to change the way we live and work has only just been scratched. Add the fundamental change to human society's ability to manipulate nature represented by the information revolution embodied in gene mapping—another technology whose application is only in its infancy—and the first half of the 21st century looks to be one of monumental changes.

Globalisation sometimes looks like chaos. Computer programs and electronic communication mean that vast quantities of the world's savings move around the globe on the basis of pre-programmed trigger mechanisms. Governments seem incapable of controlling these movements—the Mexican and the South East Asian economic crashes in the 1990s are testimony to the devastating effects that the unfettered movements of capital can have. The Internet is full of pornography, quackery, racism, misinformation and fraud that no one is apparently capable of regulating. Books are written in the United Kingdom for publishers in the Netherlands that have the copy editing done in Trinidad and the printing in Hong Kong to be sold world-wide by an Internet company located in the United States (US). Everything but the actual delivery of the book is done electronically (of course that can also be done electronically) in ways which severely restrict the ingenuity of the taxation authorities in all those countries.

Children hack into the computers of major corporations and crash their e-commerce systems just for a lark. Head offices migrate to warm places with good golfing to run things by remote control in some 21st century reincarnation of absentee landlordism. Sheep are cloned before it is decided whether it is ethical. We are informed that the food we have just eaten was genetically modified to incorporate new genetic material from other organisms and we were not even asked if we wanted them—we are not even sure what genetically modified means. Public sources are little help. A quick search of the Internet yields 75,000 hits that all tell different stories.

All of these represent change. For the most part they do not represent chaos (Kerr, 2003a). The Internet and biotechnology currently represent new technologies with very high transaction costs for consumers—whether monitoring what their children are consuming while surfing the Internet or determining what actions they need to take to ensure they are consuming wholesome food. Detection of cheaters is costly and the legal processes are outstripped by the technology. Countries fret over the long-run effect of technological and head office brain drains.

Change means that people have to alter the way that they do things. While some individuals, commonly denoted as entrepreneurs, perceive change in terms of opportunities, others find aspects of it unsettling. No matter how much government would like us to be the former, most people fit into the latter category. One major facet of government policy is to make change palatable for those who tend to see it in terms of costs rather than opportunities. The trick for policy makers is to provide a sufficient degree of order (the opposite and the antidote for chaos) to satisfy those who are made uncomfortable by change without stifling the ability of entrepreneurs to advance society's well being.

The identification of the implications of transformative technological change and the search for new mechanisms and institutions to manage and control those processes are the focus of this book.

Liability and Agriculture

The global agri-food industry has reoriented in the past decade around technological change and innovation. Both farmers and the rest of the agri-food supply chain have recognised that the long-term threat to their livelihoods is other local and regional demand for land, labour and capital. Ultimately, the agri-food sector must deliver productivity gains at least equal to other domestic sectors, or mobile resources will be bid away.

While the technological imperative is not necessarily a new feature—waves of change involving machinery (1930-60) and chemicals (1950-90) have swept through the industry in the past—the acceleration of genetically based innovation since 1985 has fundamentally challenged the industry. In the first instance, governments have encouraged the search for new technologies and products with new monopoly intellectual property rights (both patents and plant breeders' rights) and by new or different forms of government subsidy and support. The scale and complexity of using this globalised science has precipitated collaborations between traditional competitors and between public and private research organisations and has forced researchers to go beyond their borders for new science, which has worked to create barriers to new market entrants. Furthermore, the results of the research—both technology and agri-food products—has been exploited in narrow monopolistic and oligopolistic situations, which on the face of it has the potential to reduce the social benefits of these investments.

The scale, scope and speed of this transformation in the global agri-food system has generated great uncertainty in significant segments of the food production and distribution system.

The 2001 New Zealand Royal Commission that examined the issue of genetic modification as it applied to that country, included a chapter on liability, opening the discussion with the following questions: "Who is, or is not, liable for damage caused by genetic modification? Who should be? To what extent?" (p. 311).

This is a very strong statement in that it presupposes that damages from genetic modification are a foregone conclusion. This statement suggests that damage from a genetically modified plant, animal or microbe is a given—there is to be no debate on that point.

Similarly, the United Kingdom (UK) Agriculture and Environment Biotechnology Commission (2003) released a report on co-existence and liability relating to the production of genetically modified (GM) crops in the UK and recommended to the UK government that the UK Environmental Protection Act of 1990 be amended to provide financial compensation to those harmed by the commercial release of GM crops, "… irrespective of criminal liability" (p. 11).

Liability is an evolving concept, especially as it pertains to agriculture. Historically, lawsuits in crop agriculture have been mostly about production externalities, such as aerial spraying. Occasionally, an aerial application of a chemical would be too close to a neighbouring farmer's land and it would drift onto a crop belonging to another farmer. Depending on the crop, the damage could be substantial. In some instances, the farmer whose crop was adversely affected sued the commercial sprayer of the chemical for damages suffered. Another commonly cited example is the situation where a scrub bull escapes an enclosure and impregnates pure-bred cattle indiscriminately.

The genetic modification of crops has changed the nature of the liability debate and the application of the term. The commercial release of transgenic crops has created a split within agriculture, not only between countries, but within countries as well. Internationally, there has been a split between European Union (EU) countries and North America (the US and Canada). The EU views transgenic crops as a liability and will not allow domestic production of transgenic crops for large-scale food consumption, or the importing of transgenic raw materials or processed food products. North America has approved the commercial release of a variety of transgenic food crops, which, by some estimates, are incorporated into nearly 70% of all processed foods. In North America, the production of transgenic crops and the consumption of the resulting food products have become the norm. In 2003, over 80% of all soybeans grown in the US were transgenic, as were over 70% of all cotton grown. In Canada, over 65% of all canola grown was transgenic in 2003 (87% of canola grown was herbicide tolerant). Even the adoption of transgenic maize had grown rapidly, with transgenic varieties accounting for 40% of all maize grown in the US in 2003.

China and Argentina have also had long histories of commercial production of transgenic crops. Several other countries have recently begun to commercialise GM varieties, but with increased regulations. Countries such as Australia, South Africa and India are allowing for initial commercial production of some types of transgenic crops.

The split in agriculture can also be observed within domestic markets. In North America, there is a small organic agriculture market that is strongly opposed to further commercialisation of transgenic crops due to the potential for co-mingling. The organic market's fear is that, if transgenic seeds are detected in organic shipments, it will destroy organic export markets. Other producers and processors have adopted quality control systems to differentiate between GM and GM-free produce. Even in the EU, there are small amounts of transgenic crops being produced. Spain, for example, has produced between 45,000 and 55,000 acres of *Bacillus thuringiensis* (*Bt*) maize annually for the past five years (Brookes, 2002). Clearly, there are groups of producers within Europe that would adopt the technology of transgenic crops if they were allowed to do so without facing daunting market access restrictions.

This split within agriculture at both the international and domestic level gives rise to potential liability. International trade could potentially be damaged should a commodity export be tested and found to contain unacceptable levels for transgenic varieties. Domestic production could also be potentially affected by the wide-spread adoption of transgenic crop varieties. Ultimately, one overriding issue is beginning to emerge: is there a liability if a sales market is damaged by co-mingling of genetically modified seeds and, if so, who is liable?

The focus of this book is to examine the application of liability to agriculture and more specifically to agricultural biotechnology. To date, liability has not been an issue of major importance or concern to producers, firms or regulatory agencies. That is changing. This book is a preliminary attempt to examine the wide application of liability to agricultural biotechnology and discuss some of the issues that will arise as a result of this new concern.

How Did We Get Here?

The birth of modern genetics took place 50 years ago with the discovery of the double helix in deoxyribonucleic acid (DNA) by Watson and Crick. This was a substantial news story and received considerable press coverage. As with many discoveries, the vast future potential of the double helix was, to a large degree, grossly under-estimated. The research in this new field continued for the next two decades with minimal public awareness. Thirty years ago, in a California laboratory, the next major innovation in this field of genetics occurred. The extraction and insertion of genes within the genetic code of an organism was accomplished for the first time and the technology known as recombinant DNA (rDNA) was born.

This time the research community had a greater level of understanding about future applications of the technology and a heightened level of concern existed among researchers. There was so much concern that a research moratorium was enacted in 1975 to enable more to be learned about the technology of gene splicing, including the safety of those working in the laboratories. Protecting confidential information was not of prime concern—this process was very transparent, including representatives from the media, scientific magazines and the US federal government. While the issue received press coverage in scientific magazines, it was not viewed by the popular press as a major newsworthy event. Events in society forced scientific stories to the margins of the popular press—change was happening quickly in many sectors of society and the end result was the marginalisation of science.

The reason the voluntary moratorium process was transparent and open to public scrutiny was simple—it was designed to reassure those concerned that appropriate steps were being taken to prevent any actual or hypothetical risks from being realised. This process worked well as the public was informed with scientific facts of advances in the safety of the technology and the scientists of the National Institutes of Health in the US were involved in developing containment standards for proposed research projects regarding viruses and bacteria that could be harmful to humans if widespread exposure occurred.

Research involving genetically engineered viruses and bacteria continued through the 1970s and, by the end of that decade, researchers were beginning to search for ways to apply this technology to plants. Conducting genetic engineering research with plants was more time consuming than working with viruses and bacteria because the genetic code of plants was more complex. The first genetic modification of a plant occurred in 1983, but the research continued rapidly, such that by the end of the decade field trials were already underway with new crop varieties. Following several years of crop trials, a number of these new genetically modified plants were commercialised in the early and mid 1990s.

The first commercial planting of a GM crop occurred in China in 1992 (James and Krattiger, 1996). This initial planting involved 100 acres of transgenic tobacco and was done for the purpose of seed multiplication. The first commercial acreage of a GM crop for food purposes occurred in 1994—this was in the US by Calgene, with their transgenic, delayed-ripening tomato. The variety, known commercially as FlavrSavr™, was initially produced on an estimated 10,000 acres. In 1995, other crop types were introduced, including cotton, canola, potatoes and maize. James (2002) estimates that the global production of transgenic crops in 2002 was 58 million hectares.

The number of crop kinds that have been genetically modified continues to grow as an increasing number of transgenic fruits, vegetables, spices and flowers are being granted regulatory approval. Many of the new transgenic crop varieties are facing increasingly rigid regulatory standards prior to receiving variety approval. Many of these new regulations attempt to provide a clearer perspective of the risk related to the commercialisation of the prospective new transgenic crop variety.

As with any new technology, the rate and extent of development, adaptation, adoption and use have varied widely. We see five interlocking worlds where biotechnology is both creating benefits and generating potential socio-economic liabilities. Most of the focus is on the two major combatants—the US and EU— who are both major investors in and owners of the intellectual property underlying the product markets, have large, wealthy and discriminating domestic markets for differentiated food products, and have economic and geopolitical interests both in GM technologies and the global agri-food industry. While this conflict is critical to the evolution of the industry, it has already been well researched (e.g. Isaac, 2002). Our focus is more global in nature. In addition to the interests and roles of the two superpowers, we are also concerned about how the introduction of agricultural biotechnologies has influenced exporters and importers of agri-food products in other parts of the world. In that context, there are at least three definable constituencies. First, there are almost 20 major agri-food producers and exporters (beyond the US and EU) which have contributed to the development and/or adapted and adopted the technology in one or more of their export product markets. Many of these countries are members of the Cairns Group of agri-food exporters (see Chapter 4 for more discussion of their role). At the same time, there is a group of largely wealthy agri-food importing nations, mostly members of the Organisation for Economic Co-operation and Development (OECD), which have a variety of concerns about GM foods. Given that these are significant markets for agri-food products, it is inevitable that they will have a major say in how international regulations and product markets evolve. Finally, there is a large concessionary food market that is both interested in the material benefits of greater and more secure supplies of food but is also wary of new technologies that are hyped as solutions to world hunger. Ultimately, socio-economic liabilities can and do arise when developments in any of these different worlds diverge.

Differentiating Between Risk and Liability

While there are numerous options available to assess, manage and communicate risks, there are presently no obvious structures in place to manage any resulting new liabilities. Methods to determine risk, developed over the past 50 years, now involve a number of approaches, including the Risk Analysis Framework and the Precautionary Principle. There are abundant sources of literature on risk (Shrader-Frechette, 1990; Sandman, 1994; Stanbury, 2000; and Leiss, 2001) and this book will not attempt to delve into the topic of risk in any great detail. Instead, we focus our efforts on defining why liability is different from risk within the agricultural biotechnology setting, and examining how existing systems can be used to manage any resulting liabilities.

The focus of all risk analysis is to try and determine the likelihood of a risk event occurring. Risk assessment, risk management and risk communication all involve probabilistic assessment of theoretical options and evaluation of ways to minimise the potential for the development of risks, the efficient and economical containment of risks should they occur; and strategies for informing society.

Once a risk has been actualised, it is no longer a risk—it becomes a liability. Those involved in dealing with the situation are no longer dealing with probabilities, they are dealing with certainty. Once a liability is realised, it is going to cost money to rectify the situation. This is the defining difference between risk and liability. Risk is costless—liability is not. Risk is the probabilistic likelihood of an unplanned, undesired or unwanted event actually happening. This probability may range from almost perfect certainty (i.e. 100%) to infinitesimally small amounts. Once a risk has been realised, however, the probability of being affected is effectively 100%. The challenge for those involved then is not how to assess or avoid it but how to inform people about it, how to limit the impact and how to apportion the costs.

The issue of who pays in a case of liability is at the heart of many debates within biotechnology, especially agricultural biotechnology. In discussing a regulatory framework for liability, the New Zealand Royal Commission on Genetic Modification (2001) advocated that:

> ... legislation regulating genetic modification should include provision for liability and compensation; there ought to be strict liability for environmental and economic damage; "liability funds" should be established; and users of genetic modification technology should be required to give bonds for cleaning up adverse environmental effects. (p. 313)

The first of these considerations is simply saying that liability should be formally recognised and lawsuits for compensation should be allowed. The third consideration is similar to the 'superfund' concept for nuclear and chemical sites in the United States, where a pool of resources would be available to clean up any problems. Both the legislation and liability funds are relatively straightforward.

The application of the concept of strict liability for environmental and economic damages resulting from the use of transgenic crops should be a concern to all involved in the agricultural biotechnology industry. In discussing strict liability, the New Zealand Royal Commission argued:

> The rule [of strict liability] applies to the "escape" from the defendant's land of something likely to cause damage. Liability applies even if the defendant was not at fault or took all reasonable precautions to prevent the escape; the defendant must be in possession or control of the land from which the "harm" came and be making a "non-natural" use of the land; and the possibility of escape and the consequent harm must have been foreseeable, although the manner or immediate cause of the escape need not have been foreseeable. (p. 318)

The possibility of being found strictly liable in spite of enacting all possible preventative measures exposes the use of transgenic crops to the easy invitation of malicious lawsuits launched on behalf of those opposed to the concept of transgenic crops. Simply the threat of having a lawsuit brought against the uses of the technology will be enough to dramatically reduce the rate of adoption, possibly to the point of making adoption no longer commercially or economically viable. This would give birth to the largely undesirable scenario of having critics of innovation and change being in charge of commercialisation of new technologies simply through their ability to threaten to launch malicious lawsuits under strict liability. The other wording that is potentially troublesome is the reference to 'non-natural' use of the land. In plant agriculture it is difficult to fathom how this term can or would be applied by a court of law. Given that modern, industrial agriculture has been using modern improved crop varieties, sophisticated machinery and agronomic practices and chemicals, fertilisers and pesticides to improve yields and quality for more than 50 years now, it is easily arguable that conventional agriculture (and possibly even organic agriculture) could now be viewed as a 'non-natural' use of the land. Defining what is and what is not 'non-natural' use of agricultural farm land (even deciding the body that will operationalise the definition) will be hotly contested in all societies.

The fourth consideration, that users of the technology should have to post bonds to cover environmental damages, is at the heart of assessing where the liability lies. The New Zealand Royal Commission believes that the liability lies with the actual producers that use the technology of transgenic crops. However, if the federal regulatory agency has reviewed the submissions of the seed development company seeking to sell a transgenic crop and approves the transgenic variety as safe for all forms of use and consumption, it is difficult to understand how the end users can be held liable. If the use of a transgenic crop results in an environmental problem, the technical assessment required to determine this is well beyond the ability of the local producer prior to purchasing and planting the transgenic crop. This type of environmental liability would seem to result from a lack of due diligence by the federal regulatory body, whose responsibility it is to approve new crop types safe for all uses and, therefore, the liability should lie with the federal regulators.

Blanket statements about the application of strict liability to the production of agriculture can be seen as a lack of clear and focused discussion regarding all parameters of the debate. Simply stating that strict liability should apply to all environmental and economic damage resulting from transgenic crops indicates that a serious intellectual discourse has not taken place.

Study Methodology

This book frames the analysis of liability in an institutional setting. Generally, institutions are mechanisms or a bundle of rules through which choices are made and conflicts resolved (Atkinson, 1993). The formal structures of institutions are transparent and understandable but it is the informal structures that present challenges. Informal structures are operating rules, norms and cultural values that can not be easily identified at a cursory glance. This section outlines the institutional approach that we have used in the preparation of this book.

North (1990) addresses technical and institutional change and attempts to understand the commonalities and differences between the two. North believes that institutional changes are the more complex of the two due to the multifaceted interrelationships that exist with both formal and informal constraints. North argues that stakeholders have varying degrees of vested interests in institutional change and will try to influence institutional changes towards their favour.

Particular institutions tend to be best suited to govern particular circumstances. To illustrate this, we have modified Picciotto (1995). Picciotto examines some of the essential institutional fundamentals required for successful international development projects from the perspective of work undertaken by the World Bank. Picciotto uses examples of development projects to describe the institutional structures and level of operation required to ensure that resources devoted to projects are utilised in the most efficient manner possible. Specifically, he focuses on the public, private and voluntary sectors and the interaction between these sectors. This institutional methodology (Figure 1.1) provides insights into how these three stakeholders need to come together to foster successful development projects.

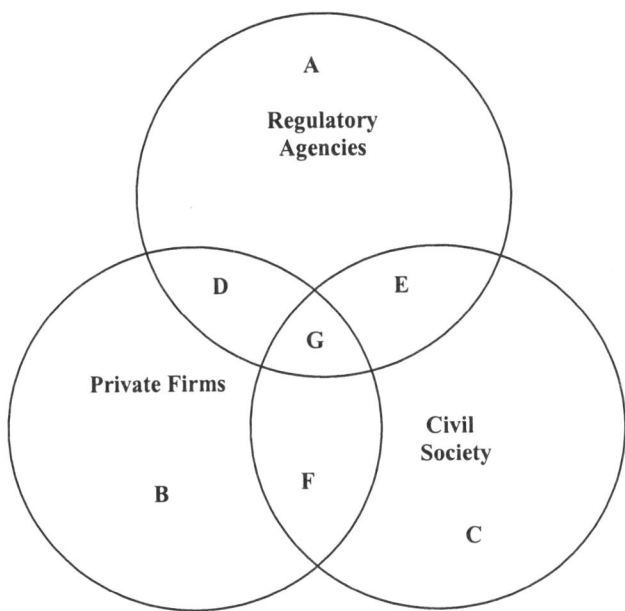

Figure 1.1: Institutional actors in biotechnology.
Source: Adapted by authors from Picciotto (1995).

Each sector represents different individuals, involves different incentives and is effective in producing goods or attributes with specific characteristics. The government sector produces public goods (A) (e.g. public health and safety), usually characterised by low excludability,[1] low rivalry[2] and low voice[3] that are involuntarily consumed by all citizens equally. Conversely, the private sector provides market goods (B) (e.g. proprietary products) that exhibit high excludability, high rivalry and low voice, and are consumed voluntarily by individuals. In contrast, the participation sector specialises in common pool goods (C) (e.g. standards that go beyond regulations but involve more than one firm), with low excludability, from low to high rivalry and high voice. The participation sector involves those who voluntarily join to obtain the benefits of collective action (Olson, 1965) and co-ordinates these collective or toll goods through some membership-based structure.

While public, private and common pool goods are important, Picciotto focuses on the relationships that develop in the three areas of overlap. The overlap between government and private goods (D) are deemed to be toll goods and are produced by public or regulated private corporations, such as public utilities. The overlap between common pool and public goods (E) are community goods. Institutionally, this is represented by hybrid organisations that are responsible for issues like rural roads. The overlap between private and common pool goods (F) are civil goods. Institutions in this category are non-governmental organisations (NGOs) and examples can be public advocacy, professional standards and civic action groups.

While Picciotto provides a good model for the assessment of resource utilisation, there are two omissions in this model that relate to managing liabilities. The first is that there is no consideration of liability management in Picciotto's model. This is important as institutional overlap inevitably creates regulatory gaps and it is the management of the related liability that is crucial to the eventual success of the technology innovation. The second oversight is Picciotto's unexplained lack of focus on the centre of his model, where all three spheres overlap (G). While all the institutions that play a role in the overlaps identified as D, E and F are thoroughly explained, he offers no insight as to what individuals and institutions are affected in this core area.

This model has been used in the development of this book to examine various issues of liability management and the role of institutions. It provides a mechanism to determine what institutions have played a role regarding a particular liability and to assess the actions taken by each institution.

---

[1] Excludability is a circumstance where individual consumers can be excluded without incurring substantial cost.

[2] Non-rival, or low subtractable, goods are ones where the consumption by one person does not diminish the ability of other persons to benefit from the good.

[3] Voice is the ability of members in a sector to have their opinion heard by those who make decisions.

Structure and Scope of the Book

The remainder of this study examines liabilities from an institutional perspective. The study examines an array of institutions—public and private, collective and hybrid—focusing on their roles, mandates and impacts on liabilities and technologies.

Chapter 2 provides a closer examination of liability and agriculture. This chapter discusses existing liability law and examines how these laws are, or have been, applied to agriculture. The focus is on civil liability and discusses the application of issues such as: negligence; strict liability; nuisance; trespass; and pollution. The discussion identifies existing case law that has been or can be applied to the analysis of these topics. The chapter offers a definition and a context for socio-economic liability, a new category of liability.

Chapter 3 provides a discussion of risk and focuses on the evolution of scientific paradigms. This discussion of knowns and unknowns is extended to policy and liability implications to demonstrate the effect on institutional actors.

Chapter 4 focuses on the consumer and the response of consumers to GM food products. While many consumers are calling for mandatory labelling for GM food products, this chapter compares what consumers have expressed in public opinion polls with what consumers have demonstrated they are willing to pay for the labelling of GM foods.

Chapter 5 examines the institutional challenges of understanding risk analysis resulting from transformative technologies and provides some comments on the difficulties associated with managing liabilities. Any time a transformative technological change occurs, such as the change that has occurred with the introduction of transgenic crops, decisions have to be made about whether the technology has any risks associated with it that are greater or less than the risks of the previous technology. The bodies or agencies that make these initial decisions have to be very careful in their assessments.

Chapter 6 discusses the relationship between liability and international trade bodies. International trade is one area that can be heavily affected by the development of liabilities. To ensure that the commercialisation of transgenic crops in one nation does not affect the ability of that nation to conduct international trade in other products and markets requires strong international governance institutions. This chapter concludes with a thorough comparison of the World Trade Organisation and the Biosafety Protocol and their respective abilities to accommodate and resolve liability within the context of international trade.

Chapter 7 offers an in-depth look at the issue of gene and pollen flow from transgenic plants and evaluates seed sterility technologies that could be used to prevent the development of liabilities. Cross-pollination from a transgenic crop to a conventional or organic crop or a weedy relative is a concern with some of the transgenic crops in production at the present time. Control of volunteers the following crop season can also be a challenge in some situations.

Chapter 8 offers, as a case study, a detailed analysis of the process that the Canadian canola industry went through to control the development of liabilities resulting from the production of transgenic canola. The initial introduction of transgenic crops in the mid 1990s was not without challenges. In Canada, the first commercial seeding of transgenic canola was in 1995. Since then, more than 12 different novel traits have been introduced. The commercial production of transgenic canola resulted in the development of some economic liabilities and the Canadian canola industry was forced to develop a variety of identity preservation and segregation systems for both conventional and transgenic canola varieties to ensure the maintenance of Canada's traditional canola markets.

Chapter 9 identifies the product differentiation alternatives that are available to deliver products to the marketplace. Part of the reason liabilities occur is due to a lack of effort on behalf of those involved in the production and processing of commodities. Each of the three major systems identified—identity preservation, segregation, and traceability—have different drivers and incentive mechanisms. Understanding why each of these systems works differently is critical to understanding how to manage the liabilities related to production and processing of transgenic crops.

Chapter 10 studies the optimal structure for regulating the production of plants that are used as a basis for pharmaceuticals and discusses some of the liabilities that are beginning to arise in this field. Drug companies are beginning to use plants to produce human proteins and antibodies required in meeting the therapeutic needs of a rapidly aging population. Concerns abound about the possibility of drug proteins that are being produced in plants becoming co-mingled with other like-variety plants and ending up being part of the ingredients consumed by an unaware public.

The final chapter offers some insights and observations about where the issue of liability and agriculture biotechnology is likely to head in the future. Ultimately, the chapter suggests that a variety of institutional innovations may be required.

Part II:

Diagnosis

Chapter Two:

Innovation and Liability

Introduction

Innovation offers the promise of a better tomorrow. Looking around our home or office, we see innovations of the past decade or two at work. Occasionally, though, innovations are used or applied in situations that the inventor would never have imagined. One only has to recall the mass hysteria of Orson Welles' radio play *War of the Worlds* to recognise the social harm that could be created by something as simple and peaceful as a radio.

Innovation, however, is not without cause and effect. Computers and robotics have drastically changed the way in which vehicles are produced, including the elimination of thousands of jobs. Ultimately, innovations force an economic reallocation of resources, both physical and human, within an economy. In fact, one is hard pressed to think of an innovation that is without any negative or adverse effect.

Historically, the majority of innovations have been concerned with inanimate objects. Innovations in construction, electronics and transportation all provide classic examples. Innovations in these fields sometimes led to social disruptions in the societies of the time, but for the most part if consumers chose to avoid the innovation, they would not be economically disadvantaged. Technological advances have recently created the ability to work innovatively with animate objects, such as plants and animals that are consumed as a regular part of our daily diets. Consumers that choose to avoid innovations in food products can be economically disadvantaged when there is no feature to distinguish between old and new products. Faced with incomplete information, consumers must either forgo the consumption of the product entirely or consume a product that may contain materials or have been produced using a technology that they consider to be inferior. As a result, they may devalue the entire product category (Gaisford *et al.*, 2001).

The maturity of an innovation reflects social comfort with any corresponding liability. Societies accept the liability of death from car accidents because the innovation of the motor car is roughly 100 years old. In other words, consumers are willing to accept the possibility of dying as the result of a car accident in exchange for access to easy, personal transportation.

The use of transgenic plants is a scant decade old and, as a result, society has not defined the comfort zone for liabilities that are arising and that may arise in the future. The present array of transgenic technologies is but the trickle prior to the deluge that may be coming. If the anticipated deluge of transgenic plant and animal innovations is to have any commercial potential, the present liabilities (both real and hypothetical) have to be satisfactorily addressed by all stakeholders.

This chapter presents an analysis of the varying tort applications that have been and are being sought against agricultural biotechnology firms. The variety of claims of liability have been summarised from various court cases that have already been heard and from those that are presently before the courts. We will not engage in a detailed examination of legal arguments related to the findings from these various legal cases, but rather attempt to provide an overview of the applications of liability pertaining to biotechnology and provide some insight into the suggested claims of liability.

Liability and Present Day Law

Legally, a liability results when an obligation is not fulfilled. From this legal perspective, there are only two kinds of liability—criminal and civil. Criminal liability occurs when there has been a criminal act committed, i.e. where someone breaks the law of the land. Criminal liability in relation to agricultural biotechnology is not the key issue, as no firm or individual has been tried for a criminal infraction.

Civil liability arises when an obligation has not been met by a party and can result in litigation for compensation on behalf of those affected. Lawsuits from those affected by thalidomide and silicone breast implants are examples of civil liability. The remainder of this section examines the areas of civil liability law that have been applied to agricultural biotechnology in several court cases. Some of the cases have already been dealt with by the courts and some are still presently in front of the courts.

Varying Statements of Claim filed by plaintiffs have sought damages for negligence, strict liability, nuisance, trespass, and pollution. Various lawsuits have also accused seed development firms of failing to conduct environmental impact assessments prior to the release of transgenic crops. This section examines each of the above concepts and claims.

Negligence

In negligence law, defendants are not responsible for every consequence of a negligent act. In other words, there are limitations on the impacts of a negligent act. The tort of negligence has three key components: the negligent act; causation; and damages. Proof of negligent liability has to include all three components. The focus of this section will be on the first component, the negligent act, as the intent is to provide an analysis of the issue rather than delve into the causation and damages of specific biotechnology cases.

The examination of a negligent act focuses on foreseeability, duty of care and standard of care. The question of foreseeability emanates from the decision of the Privy Council in Overseas Tankship (U.K.) Ltd. v. Morts Dock and Engineering Co. Ltd., The Wagon Mound (No.1) [1961] A.C. 388 (P.C.). Foreseeability in this case was seen to be based on three separate but linkable events. First, when furnace oil was discharged by the boat *Wagon Mound* into the harbour (this case is referred to as the *Wagon Mound* case), it was foreseeable that this oil would spread. Second, it was foreseeable that if the oil spread it could be ignited as a result of some unrelated event. Third, when a fire ignited due to some welding that was taking place, it was foreseeable that property damage could be expected.

This series of events could be the justification for asking seed development firms three 'what-if' questions regarding the foreseeability of, for example, GM pollen spreading. It would be realistic to pose the following three questions. First, should officials within seed development firms or federal regulatory scientists have known that the pollen from some transgenic plants had the potential to travel great distances? Second, should these officials or scientists have been able to predict that if transgenic plants were widely adopted the pollen would be widely dispersed? Third, should these officials or scientists have known that the transgenic pollen could land in fields where it was not wanted or desired?

When faced with addressing these questions, the answers are blatantly apparent. The response to the first question can only be affirmative; most farmers know that pollen has the ability to travel great distances. The response to the second question would also be positive, as it is a logical progression from the first question in that if transgenic crops were rapidly and widely adopted, likewise would be the dispersion of pollen. It seems improbable that anyone could argue anything but yes to the third question, as it would be physically impossible to prevent the transgenic pollen from landing in or on other fields once it had been released.

Based on the above, the biotechnology industry would be wise to concede foreseeability to plaintiffs. This would then shift the key focus of the debate to the issues of duty of care and standard of care. However, the concession of foreseeability would imply that there is a prima facie duty of care owed to the farming community at large. One could successfully argue that there were sufficient data available at the time of variety approval to justify foreseeability.

The focus would then examine whether there were sufficient conditions to establish a duty of care. Osborne (2000) offers a standard definition of what constitutes duty of care. Duty of care is described as "… a question of law which requires the judge to determine if the defendant is under a legal obligation to exercise reasonable care in favour of the plaintiff' (Ch. 2, p. 1). This definition of duty of care is very broad and open to varying interpretations.

The application of duty of care ultimately focuses on two key factors: Who was harmed and what is the nature of the relationship between the party suffering harm and the party causing harm? Those suffering harm can range from a single individual, as in the case of Donaghue v. Stevenson [1932] A.C. 562 (H.L.), or it can be a large group of people, such as the women that suffered from faulty silicone breast implants. While defining who was harmed as it relates to duty of care is rarely a contentious issue, determining the nature of the relationship is frequently a contentious issue.

In the case of Donaghue v. Stevenson the nature of the relationship was the contentious issue. The facts of this case from the early 1930s are that Mrs. Donaghue and a friend went to a restaurant, where her friend purchased two bottles of ginger beer. The bottles were opaque and Mrs. Donaghue poured some of her beer into a glass and drank this portion of the beer. She then poured the remainder of the bottle into the glass, at which point, the remains of a snail floated in the glass. Mrs. Donaghue became physically ill from consuming the ginger beer. The brewer of the ginger beer, Stevenson, claimed that since Mrs. Donaghue had not purchased the beer directly, there was no duty of care owed to Mrs. Donaghue. The British House of Lords disagreed with this argument and ruled that the "... manufacturer of products does owe a duty to the ultimate consumer to take reasonable care to prevent defects in its products which are likely to cause damage to a person or property" (Osborne, 2000, Ch. 1, p. 1).

More recently, the definition of duty of care has been more narrowly defined by Lord Wilberforce in Anns v. Merton London Borough Council [Note 79: [1978] A.C. 728 (H.L.)]. In this decision he states:

> First one has to ask whether, as between the alleged wrongdoer and the person who has suffered damage there is a sufficient relationship of proximity or neighbourhood such that, in the reasonable contemplation of the former, carelessness on his part may be likely to cause damage to the latter – in which case a prima facie duty of care arises. Secondly, if the first question is answered affirmatively, it is necessary to consider whether there are any considerations which ought to negative, or to reduce or limit the scope of the duty or the class of person to whom it is owed or the damages to which a breach of may give rise. (Osborne, 2000, Ch. 2, p. 2)

The use of the *Anns dictum* suggests that where there is a relationship and there is reasonable foreseeability of damages, there would be a presumption of a *prima facie* duty of care. The *Anns dictum* also created a process for a more transparent discussion of duty of care issues. Osborne (2000) believes that it was important for two additional reasons. First, it created a presumption of a duty of care in all relationships, giving rise to a reasonable foreseeability of damage to the plaintiff; and second, it placed on the defendant the unenviable and sometimes considerable burden of persuading the court why the plaintiff did not deserve to be protected from his negligent conduct.

While the *Anns dictum* originated in Britain, it is not used by British courts today. However, in Canada the *Anns dictum* has been consistently applied since the mid 1980s. Canadian courts have interpreted the first stage of the Anns test, reasonable foreseeability, as demonstrating that reasonable foreseeability of damage was probable based on the actions of the defendants. As such, there are few issues arising from this stage of the Anns test. The second stage of the Anns test, factors affecting the relationship that may restrict damages has been widely debated. Osborne believes this stage of the test:

> ... permits a full and open debate about the societal costs and benefits of recognising a duty of care. It not only allows the prima facie duty to be negated, but it also allows this duty to be restricted or modified to meet policy concerns. For example, the courts may demand that some additional element be found in the relationship, such as reliance by the plaintiff, an assumption of responsibility by the defendant, a specially close relationship, or some other element that defines the relationship more closely than foreseeability, before a duty will be recognised. (Osborne, 2000, Ch. 2, pp. 2-3)

Ultimately, the duty of care as it relates to agriculture can be analysed using two key concepts. The first concept, as it relates to the introduction of new transgenic crop varieties, would examine whether there is any reasonable foreseeability of harm that may potentially arise from the introduction of a transgenic crop variety into the environment. Secondly, the concept requires consolidation of any policy reasons or considerations that may be applied to reduce or remove the identified damage.

The first concept, reasonable foreseeability, would appear to be part of the regulatory review of the transgenic crop application. A science-based regulatory system examines all aspects for potential environmental harms. The fact that approval is granted for transgenic crop varieties would indicate that the issue of potential liabilities resulting from environmental harm has been analysed thoroughly, and the science of the day had deemed the production, processing and consumption of the crop type and the resulting food products to be safe. When a new transgenic crop receives variety approval from the federal regulatory agency, the mandate of reasonable foreseeability of harm would appear to have been satisfied. There is a lingering doubt, however, whether theoretical, hypothetical or speculative risks, which usually are not addressed in regulatory systems due to inadequate evidence, might meet the test of foreseeability.

An examination of the second concept suggests there is a need to recognise that the boundaries between the role and authority of Parliament and the role and authority of the judiciary can in some instances become clouded. Some civil society issues can become very contentious due to the debates arising among Parliamentarians and within the professional legal community. However, one issue that the courts have routinely left to the discretion of Parliament is the distribution of wealth. Technological innovations, which frequently redistribute wealth, are thus covered by policy not law.

This preference for choosing technology winners ultimately means that some group in society will lose. This concept is not new to present day innovations (the makers of buggy whips are the classic case of an industry that lost with the introduction of a new innovation, in that case, motor cars). North American society has generally chosen not to financially compensate industries or individuals that suffer financial loss due to the introduction of new innovations (see Chapter 6 for a more detailed discussion of this). Biotechnology companies can argue that they have properly followed all regulations relating to the commercialisation of transgenic crop varieties and any losses in the agriculture industry are unfortunate but not their fault. Based on previous innovations, losers in agriculture could be expected, but North American policy (and even society at large) does not call for financial remuneration to flow to those that are adversely affected by GM crops.

Those firms that have commercialised transgenic crops in North America can legitimately state that they have followed the federal regulatory process. Duty of care issues that fall into the socio-economic category may well be rejected by the judicial system given that there are no federal regulatory structures in place in North America that would justify their inclusion. Courts that are faced with tort cases claiming large financial damages due to the introduction of transgenic crop technology may decide that they do not want to rule on the re-distribution of wealth resulting from this technology and may instead suggest that this is a policy matter better dealt with in the political process.

There is ample precedence for this. Fleming (1983) notes:

> For some situations the appropriate standard of conduct is prescribed by the legislature instead of being left to the evaluative process of judge and jury. The complexity of modern life has spawned a profuse progeny of governmental regulations, demanding observance of fixed and specific precautions for the safety of industrial operations, building construction, road traffic and so forth. (p.117)

The above could remove any debate regarding the principle of standards of care. The Government of Canada, through the Canadian Food Inspection Agency (CFIA), has examined and tested transgenic crops and deemed them to be substantially equivalent to the existing varieties and has allowed their registration for commercial purposes. Given that the seed development companies have complied with the government regulations and are legally entitled to sell transgenic seed anywhere in Canada, the court may choose to recognise that any liability resulting from negligence would not be allowed in deference to policy.

Trespass and Nuisance

Some have argued cross-pollination is a form of trespass or nuisance. Making an argument claiming trespass against the production of transgenic crops would seem to be difficult. In the decision in Philips v. California Standard Co. (1960), 31 W.W.R. 331 (Alta. S.C.) the judge identifies that in England and Canada trespass involves a physical entry on the property of another. While there is no physical entry in the sense of a human or an animal entering a property, there is physical entry of GM pollen. There have been some agriculture cases of trespass involving livestock but there are no known cases of legal action based on plants, portions of plants or weeds. Existing case law suggests that trespass only pertains to entries that can be physically prevented. To be successful in arguing that GM pollen should have been controlled would mean proving that the CFIA was negligent when approving GM canola. As a result, it would be doubtful that an argument based upon trespass could succeed.

Nuisance cases may be equally problematic. There are two variations of nuisance, private and public. Private nuisance can be defined as a substantial and reasonable interference with the use and enjoyment of land. In public nuisance there has to be a defined group or class of individuals that has perceived harm at the hands of another.

While the discussions regarding nuisance are many, the decision of Lord Lloyd of Berwick in Hunter v. Canary Wharf Ltd. (1997) NLOR No. 324 NLC 197044701 provides a very concise definition of private nuisance:

> Private nuisances are of three kinds. They are (1) nuisance by encroachment on a neighbour's land; (2) nuisance by direct physical injury to a neighbour's land; and (3) nuisance by interference with a neighbour's quiet enjoyment of his land. In the case of encroachment the plaintiff may have a remedy by way of abatement. In other cases he may be entitled to an injunction. But where he claims damages, the measure of damage in cases (1) and (2) will be the diminution in the value of the land. This will usually (though not always) be equal to the cost of reinstatement. The loss resulting from diminution in the value of the land is a loss suffered by the owner or occupier with the exclusive right to possession (as the case may be) or both, since it is they alone who have a proprietary interest, or stake, in the land. (Notes 39 and 40)

An examination of the three kinds of nuisance is insightful. The first, encroachment, would only apply if it could be physically proven that GM pollen has encroached on the land of another producer. This argument would presumably follow the logic needed to prove trespass. Thus, it may be difficult to claim nuisance if there is no proof of encroachment.

The second form of nuisance, direct physical injury to a neighbour's land, may be a possible argument if the presence of GM pollen can be viewed as a direct injury. Pollen from any plant species is only viable for a few hours, thus making it challenging to demonstrate how the presence of GM canola pollen, for example, could be viewed as an injury to the land. The pollen from GM canola could be an indirect damage, as it may affect the plants growing on the land, but the pollen can not physically damage the land.

If an argument were made for physical injury, the damages awarded would be for the decrease in the value of the land or the cost of restoring the land to its original form. The latter option would only be possible with the complete ban of GM crops in Canada and the US. To show a decrease in land value would be extremely problematic, as there is no identifiable market for land that has not produced a GM crop. It would be very problematic in North America to try and determine what, if any, market premium exists for land that has not produced a GM crop.

The third kind of nuisance, interference with enjoyment of land, also seems a challenging argument, as pollen, in the conventional sense, is not an irritant about which individuals complain. Public nuisance is more difficult to prove than private nuisance, as proof of special damage must be made. Successful public nuisance cases have occurred where some form of toxic material is released by a company into a local body of water and poisons most or all of the local aquatic wildlife, thus being a public nuisance to local fishermen. A caveat to this is if 'interference with enjoyment of the land' can be interpreted to have a financial meaning. If a court decides that a farmer's financial enjoyment has been affected, the importance of nuisance may increase.

Furthermore, individual farmers are also often protected from nuisance suits. In Canada, for example, the Agriculture Operations Act (1995) of Saskatchewan states:

> The owner or operator of an agricultural operation is not liable to any person in nuisance with respect to the carrying on of the agricultural operation, and may not be prevented by injunction or other order of any court from carrying on the agricultural operation on the grounds of nuisance where the owner or operator uses normally accepted agricultural practices with respect to the agricultural operations. (p. 5)

A very important issue that develops from the above quote is how much time has to pass before a new technology is 'consistent with accepted customs and standards'. Two cases of aerial crop spraying may shed light on the issue. In the case of Mihalchuk v. Ratke (1966) 57 D.L.R. (2d) 269 a lawsuit arose from an aerial spray application made in 1965 and the judge noted that this was an unusual operation. The second case, Cruise v. Niessen (1977) 82 D.L.R. (3d) 190 is based on an aerial application made in 1975 and the judge noted that aerial spraying is no longer viewed as unusual. In the case of aerial spraying, ten years was enough to make this technology a common practice. Including the crop produced in 2004, genetically modified crops have now been grown for ten years. The transgenic canola acreage for 2003 was between 65% and 70% of all canola acres in the province. Two out of three canola growers use this crop technology and it has become a very popular option for weed management practices.

Strict Liability

The issue of strict liability is also relevant to this case. Typically, strict liabilities are found for one-time occurrences, such as in the case of Rylands v. Fletcher (1868), L.R. 3 H.L. 330, affg L.R. 1 Ex. 265. The ruling in Rylands v. Fletcher is based on the premise that an item or product not naturally occurring, but being stored on one's land, may be inherently dangerous. In this case from Britain, Fletcher owned a mill and he constructed a reservoir to supply it with water. During the construction of the reservoir, the contractors discovered five long-abandoned vertical shafts. Not knowing they were abandoned mineshafts, the shafts were filled with soil. The reservoir was partially filled with water and shortly afterwards one of the soil-filled shafts gave way, flooding the nearby coal mine owned by Rylands. Rylands sued Fletcher for the destruction of his mine. The ruling in Rylands v. Fletcher describes the item or product being stored on one's land as not naturally occurring and therefore, this product is inherently dangerous. In this case, a large inland body of water was viewed as a dangerous activity. This ruling has three important considerations for any GM cases. First, the drift of GM pollen is not a one-time occurrence, rather it happens annually for a period of 3-6 weeks. Secondly, it would seem impossible to argue that GM pollen is stored in any form or fashion upon a farm. Finally, there are no scientific arguments that can be made in favour of the GM pollen being inherently dangerous, given that regulatory agencies have approved these varieties for all uses in both Canada and the US.

Pollution

Some have claimed that GM pollen is a form of pollution as defined under environmental protection legislation. Most existing pollution legislation refers to the discharge of a pollutant, defining the obligations of an owner of the pollutant and of those who control a pollutant that causes the damage or loss. It is important to note that legislation of this kind refers to the discharge of hazardous substances. The requirements for a product to be listed as a hazardous substance are that the product has the potential to harm the environment, human health and/or other living organisms. Such substances are usually listed in some form in a related environmental act.

Discharge as it relates to GM crops will vary, depending on whether it is applied to the GM seed or the GM pollen. There is no doubt that GM seeds have been deliberately discharged into the environment by seed development companies after receiving approval for unconditional release by federal regulators. In North America, the federal government approved the release (and thereby the discharge) of GM crops. Discharge as it relates to the pollen is a very complex issue to assess. Pollen from GM plants is released, dispersed or emitted into the environment just like the pollen of every other plant. The difficulty of this is how could a farmer who has been negatively affected identify which field was responsible for dispersing GM pollen.

Both the 'owner of a pollutant' and the 'person having control of a pollutant' prior to first discharge include the successor, assignee, executor or administrator of the owner or person. Executor and administrator are not applicable in this case by their own definition. One issue of importance is whether the use of a Technical Use Agreement (TUA) between the seed provider and the farmer identifies the farmer that signs the TUA as either a successor or assignee. Black's Law Dictionary (1979) defines assignment as, "a transfer or making over to another of the whole of any property, real or personal, in possession or in action, or of any estate or rights therein. ... the transfer by a party of all of its rights to some kind of property, usually intangible property such as rights in a lease, mortgage, agreement of sale or a partnership" (p.109). Based on this definition, a TUA would not be an assignment of property. Rather, it is an agreement that allows the farmer to use the technology in exchange for a fee, and hence remains the property of the technology owner. Arguably, the enforcement of intellectual property rights embedded in patents and plant breeders rights would similarly vest ownership with the technology provider.

Pollutant and pollution and their relation to GM pollen will be crucial to the success of any potential lawsuit. We do know that GM seed and pollen are not currently listed as hazardous substances in any country and there are no published data to suggest that consumption of these products is physically harmful. The organic industry argues that the existence of GM pollen is harmful to their agricultural use of the environment. However, it must be made very clear that harm as it is applied in the organic sense is not harm in the physical sense of the word—the application of the word harm in this case is financial. The question courts will be asked to decide upon is whether harm as it is applied within the definition of pollution extends to financial harm.

Some opponents of agricultural biotechnology have tried to advance the argument that GM crops should have been subjected to more extensive environmental assessments by all levels of government, be it federal and state or provincial. The regulation and approval of plants is almost exclusively a national responsibility—no sub-national government has yet developed legislation regarding the regulation of plants and many are constitutionally prevented from doing so.

What is Socio-Economic Liability?

For our purposes, a socio-economic liability is defined as the decline in social trust for all innovations and the economic decline from commercialisation delays when a company or government regulatory body fails to meet their publicly stated objectives, which are, ultimately, their social responsibility.

The differentiating feature between socio-economic liabilities and conventional legal applications of liability is that, with socio-economic liabilities, there is often no direct identifiable failure. Because there does not need to be an identifiable failure, there is no possibility of foreseeability and, therefore, no standard or duty of care exists between the social challenge and the owner of the innovation. This form of liability has no direct causation and firms that operate well outside the area of the innovation can be negatively affected by the actions of other firms in the industry.

When a regulatory failure occurs, the resulting media coverage changes the social perception of innovation. Negative media coverage will result in some consumers that were initially indifferent to a specific innovation becoming concerned about it or possibly even opposed to innovation in general. The bottom line is there is a loss of social trust in innovation, be it specific or general. This is not to say that the decline in trust will always be long term, or that it can not be reversed, but rather that there will be fewer consumers willing to express support for an innovation or the resulting commercial products.

Regulatory failure increases the likelihood that commercialisation will be delayed, creating negative economic consequences. Heller (1995) has estimated that a one-year delay in commercialising an innovation reduces the rate of return on investment for a new product by 2.8%, while a two-year delay results in a reduction of 5.2%. A recent study of plant biotechnology by Gianessi *et al.* (2002) found that current transgenic plants increased yields by 4 billion pounds annually and reduced pesticide use by 46 million pounds annually. The combined value of these effects was estimated to be US$1.5 billion. Delays in commercialisation of transgenic plants would reduce the benefits of higher yields for producers and increase the application of pesticides in the environment. This would certainly result in producers being economically worse off and arguably those sensitive to pesticides could also be deemed to be economically worse off.

The absence of criminal or civil liabilities does not negate the fact that when a regulatory failure occurs, there are social externalities that are ultimately borne by other firms and by consumers. While these externalities may not generate a level of harm that is large or severe enough to trigger litigation for compensation, a socio-economic liability is created with regulatory failures.

To illustrate this point, it is worth while to examine how the co-mingling of StarLink™ maize with conventional maize resulted in a socio-economic liability. The detected presence of StarLink™ maize in processed maize products in the United States in September 2000, resulted in a very large liability for Aventis (Lin, *et al.*, 2003). At the time of detection, StarLink™ maize (which expressed the *Cry9c Bt* gene) had only received regulatory approval for use as animal feed. The US Environmental Protection Agency (EPA) required Aventis to establish a system to ensure that all StarLink™ maize produced be channelled into animal feed markets. Aventis established a 'Stewardship Program' to satisfy the EPA's objective but the programme failed miserably. The programme outlined the requirements of contract registration for all fields of StarLink™ maize, the use of buffer zones and separate binning of harvested maize. The process of events has been described by Harl *et al.* (2000), but what has not been done is to examine where the responsibility for the creation of this liability lies.

Initially, the biotechnology industry, government, concerned social actors and academics placed all the blame squarely on Aventis. The 'Stewardship Program' looked great on paper yet the implementation of the programme failed in two key ways. First, government and industry reports indicate that many seed distributors were not aware that this seed had to be destined for use only as feed. Second, numerous farmers stated that they never signed the required production contract, let alone followed the buffer zones and separate binning requirements specified in the 'Stewardship Program'. By all accounts, Aventis developed a strategy to contain this transgenic maize variety but failed to monitor the programme once it was implemented. The question is whether all the liability should rest with Aventis?

Our analysis of the event suggests that a portion of the liability should indeed lie with the EPA. The EPA had made split-run decisions, whereby varieties must be kept separate, in the past and no problems had arisen. In the StarLink™ case, the EPA required Aventis to develop a plan to contain the maize for the animal feed market. The plan was created, the EPA reviewed it and approved StarLink™ for animal feed use and then the EPA abandoned its responsibility to monitor and enforce the proper containment of this variety. As federal regulators, the EPA failed in the StarLink™ case to properly ensure that the 'Stewardship Program' was being followed and that those involved with the selling and growing of StarLink™ maize were properly aware of the regulatory requirements. In its role as federal regulator, the EPA placed absolute trust in Aventis. This is something that a federal regulatory body can not and should not ever do. To place a regulatory requirement upon industry and then blindly trust that this requirement would be followed without incident, indicates either extreme naivety within the EPA or a lack of appreciation for the role regulatory bodies have to play in the management of liabilities, especially those associated with new, transformative technologies such as agricultural biotechnology.

Socio-economic liabilities result when the federal regulator, private companies or both fail to recognise the marketplace impacts that may result from a regulatory failure. In the StarLink™ case, the US lost maize sales to Japan because of the adventitious presence of StarLink™. The inability of US maize exporters to guarantee their shipments were StarLink™-free enabled Brazil to export maize to Japan for the first time in 20 years. Market access for like-products can be severely jeopardised when a transgenic plant creates a socio-economic liability. In today's global marketplace, a very localised physical problem can have negative financial impacts across the globe.

Current Court Action on Liability

There is a clear delineation in the approach taken by North American governments versus that taken by the EU and the member state governments. The EU has adopted a precautionary, go-slow approach and has placed a moratorium on the commercialisation of transgenic crop varieties. North American governments have moved to legislate in favour of this technology and have approved numerous transgenic crop varieties.

Most court cases in Europe to date have dealt with the opponents of GM crops and their destruction of test plots. Meanwhile, European governments are examining their existing liability laws to determine if they adequately address potential lawsuits involving co-mingling or the unintended presence of GM crops. For instance, the recent UK study on co-existence of GM crops recommends that compensation be made available to those that suffer financial losses due to the presence of GM crops through no fault of their own (AEBC, 2003). Over the past few years there have been a variety of court cases in North America involving liability.

Given that those negatively affected by the introduction of transgenic crops have little or no recourse of action with the regulatory agencies in North America, they have indicated that they may turn to the legal system in an attempt to seek remuneration. In Canada, for instance, the Saskatchewan Organic Directorate (SOD) representing a group of organic producers have filed a lawsuit against Monsanto and Aventis seeking compensation for alleged lost organic canola markets following the commercialisation of transgenic canola. In addition to compensation for damages, the SOD is seeking an injunction against the commercialisation of GM wheat. The justification for the injunction against GM wheat is that the SOD believes that the presence of GM wheat will jeopardise their ability to export organic GM-free wheat to Europe.

One other Canadian court case is the counter-suit launched by farmer Percy Schmeiser against Monsanto for allegedly polluting his farmland. This case is separate from the patent infringement case that is presently before the Supreme Court of Canada. Schmeiser lost both the trial and an appeal at the Federal Court regarding his illegal use of Monsanto's patented canola technology but has been granted leave to appeal to the Supreme Court of Canada. Schmeiser's counter suit alleging pollution by Monsanto has been put on hold until a decision is handed down by the Supreme Court in the patent infringement case.

Liability lawsuits in agriculture related to potential market losses have the potential to severely limit future benefits of biotechnology. At the present, there is a growing unease within the agricultural biotechnology industry that legal injunctions may become increasingly common as opponents of GM crops attempt to use this strategy to prevent the commercialisation of new GM varieties.

Conclusions

Consumer trust in government and industry has declined in all industrialised countries. At one level, consumers (and others in society) want to see that, when a risk is realised due to the actions of one or more firms, there are negative impacts not only on their industry but on all industrial sectors.

The lack of theory and peer reviewed scientific evidence related to the long-term consumption and use of transgenic crops has four important implications for the relationship between law and agriculture. First, there is an absence of causal science that links one event to another event. Given that the law requires foreseeability to establish a duty and standard of care, the absence of any accepted causal evidence can be a major constraint on using the legal system to adjudicate claims made by those unhappy with the new technology.

Second, many of the instances of liability lie outside of any identifiable relationship. This means that it is virtually impossible to predict what relationships may develop between liability and damages. Consumers may see or read an article or story about a regulatory failure or violation involving an agricultural biotechnology firm or product that they do not have anything to do with, but this may still may substantially impact their food purchasing habits. Ultimately, socio-economic liabilities will involve a wide range of actors who are not part of any relationship.

Third, there is very little precedence in law for cases involving GM crops, and those cases that have been adjudicated have yielded somewhat inconsistent results. Some courts have decided to follow the science of the argument and respect the intellectual property rights of the firm involved, while others have been swayed by popular public opinion. This results in an even greater risk that litigation will be problematic.

Finally, the lack of long-term experience with this new technology contributes to a lack of comfort with the general pace and direction of innovation in the global agri-food system. Adoption of GM crop technology happened so rapidly and to such a large degree that the marketplace in many instances has hardly begun to adjust to this transformation. Any shock to an efficient system produces shockwaves that reach the edge of the system and then have the potential to rebound back to the centre of the system. The effect of the shockwaves resulting from commercialisation of GM crops is likely to continue to affect the overall agricultural marketplace for years to come.

Chapter Three:

Social Amplification of Risk

Introduction

Transformative technologies are created from the basic sciences; they are developed through a complicated process of research, development, adaptation, adoption and ultimately placed into the public hands for use. These technologies change the ways that society views the certainty and uncertainty of science.

While the food sector has long been extensively governed by industry standards and government regulations, the rules for biotechnology-based agri-food products are still evolving. The nascent institutional structure for managing risks of new products has been buffeted by high profile releases of incomplete and controversial scientific information. In response, regulatory systems have introduced expensive and often sub-optimal rules and procedures.

Certain scientific questions that have arisen do not have easy answers. Many are questioning how the instruments of science and its tools define the nature of the research. Given that unanswered query, many ask how the specifics of biotechnology affect appraisal of liabilities, development of policy related to potential liabilities, and national and global governance. These questions and their preliminary and often incomplete answers create tensions between transformative technologies, the instruments of science and their use in diagnosis of causes of liabilities, their mitigation and their management. Without understanding these tensions, we cannot hope to understand the limitations that instruments of science bring as both problems and solutions.

This chapter identifies the importance of the scientific grounding of institutions and policies and the impact of incomplete science on the management of liabilities. Complex issues and public expectations from science in relation to transformative technologies may not have generally acceptable, or even well defined, questions and answers, which compounds the problems of managing transformative technological change.

Background

Plant biotechnology is an umbrella term used for a number of interconnected disciplinary concepts, theories and techniques. It offers the promise of generating new products and processes that will make more efficient use of resources and provide food or bio-product opportunities, including functional foods and pharmaceuticals.

Two issues occupy the public agenda. First, unresolved or complicated liabilities arising from previous transformative technologies, new chemistry and nuclear energy are still embedded in public experience and memory. Biotechnology now presents changes that are generating even greater public reaction. Second, plant biotechnology as applied science creates a new generation of plants, whether for food, industrial use or plant manufactured pharmaceuticals, which will have a major impact on humans and animals directly through consumption and indirectly through pollen dispersal. Pollens, which are vehicles for transmission of genetic material, are also a source of food and nutrients for pollinating insects and for humans (most often as nutraceuticals).

The challenge facing scientists and regulators is to understand how this new technology creates risks and liabilities. As with any transformative technology, this presents a challenge of working with non-paradigmatic or evolving science. Biotechnology has destabilised our concept of what is 'normal' and will continue to do so as it evolves. Thomas Kuhn (1970) identified the conceptual tool of 'normal science' in order to illustrate how new technologies and knowledge destabilise the system and generate new controversies. This approach to scientific development, in this case the science of production of new plants, forces us to analyse basic assumptions about the role of science and the proper conduct of scientific enquiry. We can then assess how paradigmatic changes can either identify, create or enhance liabilities or, at times, reduce pre-existing liabilities.

Liabilities arise from transformative technologies because in some instances there are miscalculations during the scientific risk assessment. Risk assessment is understood to be 'a scientifically based process', but it is not clear what that might mean. The risk assessment process engages authoritative interpreters as experts who have a firm and consensual understanding of the issues that may arise with such technologies.

After all, there are only two choices, to wait for the outcome or presuppose its circumstantial occurrence and react. The source of authoritative interpretation is experience and knowledge. The latter depends on open dissemination of scientific information, which becomes more complex as disciplinary or interdisciplinary expertise fragments and specialises and as interpretative authority becomes multi-domain and often resides in amorphous groups of individuals. Citizens, whether in the corporate world, as consumers or at times as regulators, become overwhelmed. Many are inundated with issues and answers because of the fragmentation of information. In sum, public distrust of both the interpretative authorities and regulators guarantees divisions in acceptance of risk and precipitates new liabilities. The providers of scientific advice also become caught in the conundrum and at such times they respond by attempting to clarify and comfort but often only serve to offer responses that confuse and heighten concerns.

The Theory of Risk and Markets

Publicly managed risk analysis systems, based on scientific risk analysis, are vital to creating the trust necessary to launch new technologies and new products. Van den Daele *et al.* (1997) identify three types of risk that affect the safety of a product and consumer perceptions of those risks. First, probabilistic risks involve those theoretically-grounded and empirically-demonstrated risks related to the product or its technology. The methods and much of the evidence are part of Kuhn's 'normal science' and are available in peer-reviewed journals or public records (Kuhn, 1970). Second, hypothetical risks, in contrast, involve those possibilities that are grounded in accepted theory but lack empirical experience or evidence that can establish probabilities. Most of these can be identified in academic literature. Third, speculative risks, in contrast to the other two areas, have neither established theory nor experience to back them up. Those speculative risks are often identified in working papers or other developing literature. Beyond that, almost any correlations can be made to show the potential for risk, irrespective of whether there is any theoretical basis for the possibility.

New products can raise concerns about all three types of risk, which could generate criminal, legal and socio-economic liabilities. Effective risk analysis systems encompass elements that work to assess, manage and communicate the risks to consumers and citizens. Risk assessment, which involves an *ex ante* evaluation of risks and benefits of GM foods, is usually based on an objective examination of the 'normal' science on a case-by-case basis. Risk management, which involves the human system of rules and procedures, is designed to contain products during the research and commercialisation phase and undertake post-release monitoring and surveillance. Risk communication, which involves dissemination of information, attempts to narrow the gap between actual risks and the perceptions of risk.

As one might imagine, most risk analysis systems should be easily able to handle risk assessment and risk management for iterative inventions (as both relatively mechanistically use 'normal' science). Transformative technologies pose greater challenges given the hypothetical or speculative nature of much of the science. Risk communication, however, is extremely complex. Transformative technologies, like biotechnology, are especially difficult to communicate about because of the wide array of speculative issues that science cannot adequately address. In the absence of complete certainty about, or at least significant experience with, a new technology or product, risk analysis systems must inherently be based on trust. In turn, trust is instilled as a result of reactions to risk assessments that are carried out by institutions. Trust grows as risk management procedures address public concerns. Finally, trust is firmly embedded in the public via the success of the risk communication process that develops over time. Strong institutions are crucial for successful and trustworthy food safety systems.

Frequently, when food safety institutions attempt to deal with risk they adopt the approach of risk elimination, which can result in costly and, at times, unnecessary regulations and/or standards. There will always be risk associated with everything individuals do. The key is to lower the level of risk until it is within an individual's or community's comfort zone. This is the challenge when dealing with food safety. While the utmost care is taken to ensure that products on shop shelves are as safe as possible, there is still a minimum level of risk that must be faced when purchasing any foods.

Traditional risk assessment theory suggests that risk is a combined measurement of the length of exposure multiplied by the level of adverse effects of the agent to other organisms (i.e. hazard). This can be expressed in the following formula: RISK = HAZARD x EXPOSURE. If the time of exposure is brief (e.g. fractions of a second) and the level of hazard is triggered only with high doses, the level of risk could be low or minimal. Scientists have used this formula to evaluate whether initial research findings should proceed or be halted. If an assessment was conducted and the level of risk was determined to be high, then government agencies may not approve the technology or product for commercial release. Experts may well have a different view about the level of risk associated with a new product or technology than the public. As a result, experts are often confused by consumer reactions to new products and technologies.

Sandman (1994) has argued that in some cases regulators should instead use an alternative formula for understanding consumer perceptions of risk: i.e. RISK = HAZARD x OUTRAGE. Sandman believes the old formula underestimated the actual level of risk because it ignored any public response (i.e. outrage). Public concern is often focused on whether the risk is acceptable rather than on the scientific assessment of risk. This has important implications for risk communications, as food safety institutions must address consumer outrage in their response to the risk assessment.

Craven and Johnson (1999) in writing about food scares discuss the new approach that risk communicators must take when faced with consumer concerns.

> The science of risk communication is still relatively new, though valid and effective precepts have been clearly defined. Mitigating a hazard itself does not mitigate the outrage about the hazard. To defuse a scare, outrage must be addressed, that is, the public's particular concerns must be addressed and dealt with in a way responsive to their emotional needs regarding the issue. (p. 164)

The difficulty for risk communicators is to determine which scares, whether related to food or not, will provoke an outrage response from consumers. It is clear that those responsible for risk communication in the biotech industry and food safety institutions were not anticipating the consumer outrage that developed in the EU over the introduction of GM agri-food products.

Snow (1997) suggests four critical elements are necessary to deal with consumer outrage. First, the communications must be open and fully disclose the threat and issues. Second, the communicator must acknowledge responsibility for the issue. Third, the communicator must be courteous (even if the public response is angry and impolite). Fourth, the communicator must demonstrate compassion, by respectfully recognising and addressing people's fears and apprehensions.

The reason for the failure of EU, and particularly British, food safety systems to effectively address consumer fears regarding GM foods becomes clearer when their responses are evaluated in the context of these four approaches—the British and the EU's food safety institutions acted in the complete opposite manner as prescribed by Snow. Had these food safety institutions been better risk communicators, it might not have been necessary to implement costly and often ineffective regulatory and labelling regimes that have effectively delayed use of the technology for more than five years now. In contrast, the US risk communication institutions responded more to the outrage than the hazard, effectively lowering the cost of maintaining trust in the products and the regulator. Risk communication is especially difficult when scientists are forced to respond and interpret risks related to a transformative technology.

Paradigmatic Nature of Transformative Technologies

Kuhn (1970) constructed a generalised picture of the process by which a science is born and undergoes change and development. Kuhn used the term 'paradigm' to denote the body of knowledge that was part of 'normal science', the study of natural phenomena, which led to a set of theories that explained differing viewpoints. In turn such activities yield the development of a pre-paradigm (or accepted body of knowledge), which after much empirical testing, evolves into a paradigm. A paradigm thus has a number of questions and answers that are accepted as 'knowns'. The so-called 'knowns' are related to each other through research and publication. So-called 'normal science' undergoes an evolutionary process leading to the further articulation of the paradigm through the use of existing theory to predict, to solve and to develop applications. During the course of scientific enquiry, new phenomena are often discovered and new or revised explanations are developed for these phenomena. In other words, scientific questions emerge because the unexpected occurs and new answers become the objects of enquiry. Things that are 'unknown' drive the search for new theories or answers, or new 'knowns'. People discover problems not previously known, and existing theory sometimes proves unable to account for the anomalous facts. Sometimes in such circumstances a researcher can define and exemplify a new conceptual and methodological framework not commensurate with the old that leads to a new solution, thereby allowing the continuation of 'normal science' within a new paradigm (Green, 1971).

Applications to Biotechnology

Kuhn's examples of paradigms, while drawn entirely from the history of the physical sciences, have a place in the context of many transformative technologies, especially in biotechnology and food science (Khachatourians, 2001). The challenge for risk and liability management is to distinguish between situations of normal, emerging and speculative science.

The classic quadrant box (Table 3.1) can be used to describe the relationships between questions or answers that represent knowns (Ks) and unknowns (UKs). There can be only four combinations or boxes: where both the question and answer are known (K-K); where one of either the question or answer is known but the other is not (K-UK, UK-K); and finally where neither the questions nor the answers are known (UK-UK).

Table 3.1: Kuhn's paradigm of knowns and unknowns

		Answer	
		Known	Unknown
	Known	K-K	K-UK
Question			
	Unknown	UK-K	UK-UK

Source: Khachatourians (2001)

Certainties or uncertainties of agricultural biotechnology in general, and plant biotechnology as a specific case, can be assigned to any of the four quadrants. The top left quadrant (i.e. K-K) provides few difficulties for most scientists, regulators and citizens. They usually do not represent uncertainties as the science is clear, information is readily available, and the community at large agrees that both questions and answers are 'known'. Stakeholders, however, sometimes disagree on the spectrum and the scale of that which is 'known'. As a result, the lower-left and upper-right quadrants (i.e. UK-K, K-UK) represent situations where there is increasing concern about the products, and, hence, where unanticipated criminal, civil and socio-economic liabilities may arise. They occur because the interpreter—the scientist, the regulator or the activist— does not have the authority to bring certainty and make the questions or answers believable. The final situation is when neither the question nor the answer is clearly known. It is in this quadrant that we are unaware of the risks and we are overwhelmed by the questions.

Once we employ real examples to Table 3.1, we can see how concerns can escalate. This situation was dramatically illustrated by Khachatourians (2001) for the cattle industry. That example populated the quadrants with four illustrations: the provision of antibiotics in feed represented a known-known example; the recombinantly-derived bovine somatotrophin (rBST) hormone represented a case of known questions and unknown answers; the rendering practices and their relationship to catastrophic bovine spongiform encephalopathy (BSE) offered known answers but unknown questions; and the presence and transmission of human cases of the variant Creutzfeld-Jakob disease demonstrated a case of unknowns.

The quadrant model can be applied to plant biotechnology, the development of transgenics and impacts of dispersal of pollens (Table 3.2). Four scenarios of known-unknown situations arise: inter-specific pollination (ISP); cross-species pollination at short or long distances (CSP); the occurrence of pollen-induced type-1 allergic reactions (PIAR) in humans; and the impact of pollen flow from plant made pharmaceuticals (PMPs).

Table 3.2: Kuhn's paradigm of knowns and unknowns related to transgene pollination

		Answer	
		Known	Unknown
Question	Known	ISP	PIAR
	Unknown	CSP	PMPs

Source: Authors

The first case, ISP, represents scientific questions and answers, which are known. As such, liability can be easily discerned because of historical experience and fundamental knowledge of plant genetics and agronomy. Inter-specific pollination from the standpoint of plant variety production is one means for improved quality. In this case, both the questions and the answers related to risks are clearly known. There are no contentious liability issues and most questions and answers are known due to traditional plant breeding experience.

In the second and third scenarios, the CSP and PIAR cases, normal science did not expect the tractability issue of pollens to be so precisely resolvable. Contemporary tools of molecular science now permit us to establish the relationship between pollens and pollination and allergenic and human immune responses.

Cross species pollination represents a case of an unknown question and known answer. The difficulty is that while we have significant evidence of cross-pollination, it is not certain whether or not answers provided as 'known' will satisfy potential liability issues. Pollen, through pollinators (including wind) can move over variably large distances. For example, the question of whether pollen from canola plants (*Brassica rapa*) will cross-pollinate that of mustard appears to be clear. The test system for finding the answer is known although scientists continue to debate the data and the long-term effects and, hence, the liability. Canadian government regulations require either large isolation zones (200m) or

10m wide borders of synchronously flowering, non-transgenic *B. napus* to contain transgenic pollen. Staniland *et al.* (2000), however, have shown the ineffectiveness of border areas in containing transgenic *B. napus* pollen. When seed samples from out-crossed seeds were harvested from the border areas they showed out-crossing rates which averaged 0.70% at 0m which then declined exponentially to 0.03% at 30m. While more than 80% of the total outcross events detected occurred in the first 10m of border area, suggesting that border areas might effectively reduce pollen-mediated gene flow in *B. napus,* border areas did not completely eliminate it.

One solution to the liability created by CSP is to deploy transformative technologies, which can intervene with the consequence of a UK-K situation. Saeglitz *et al.* (2000) proposed the use of 'bait plants' as an effective tool to control the flow of genes through pollen-mediated gene escape in mating plants. In their study, cytoplasmically male-sterile (CMS) sugar beet plants were tested with regard to their potential for monitoring transgene escape. The transgenic marker they employed was an rDNA encoding viral (beet necrotic yellow vein virus) resistance and antibiotic (kanamycin) and herbicide (glufosinate) tolerance. In a field trial, the effectiveness of a hemp strip for deployment of a containment strategy was tested. They measured the frequency of pollinated CMS bait plants placed at different distances and directions from a transgenic pollen source. Their results demonstrated the ineffectiveness of the containment strategy. Physiological and molecular tests confirmed the escape and production of transgenic offspring more than 200m behind the hemp containment. In relation to Kuhn's model, this example demonstrates that the answer—absolute containment—is now known to be unlikely to be effective. The validity of a CMS-bait plant as a detection system or as a useful tool for other monitoring purposes remains in the domain of 'unknown'.

A subset of the CSP case would be the emergence of new transgenic hybrids because of cross-pollination over distances exceeding several kilometres. In this case the evidence affirms that cross-pollination can occur. The tools of DNA detection and use of signature DNA make it possible to arrive at a 'known' answer. However, the question of what determines the cross-pollination is in the realm of the 'unknown'. The liabilities associated with the question, while minimal if unknown, could be very large when it becomes known. Hudson *et al.* (2001) tagged the pollen-active LAT59 promoter from tomato to express a green fluorescent-protein (GFP) encoding gene (derived from the jellyfish *Aequoria victoria*) and a label of gene expression and protein localisation (from tobacco pollen). This promoter is preferentially expressed in anthers and pollen. GFP-tagged pollen was developed as a tool for tracking the movement of transgenic plant pollen. When GFP is expressed it produces a protein that fluoresces a bright green, which can be used for measuring the spatial distribution patterns of transgenic pollen in the environment. While this will help us to refine our knowledge of actual gene flow in field conditions, it does not in and of itself help to refine the theory that will provide causal predictions.

In the third case, PIAR, the technical question and associated liability question are 'known', but the answers remain unknown. There is a shortage of answers about normal allergens, never mind those containing new GM proteins. Pollens that induce significantly greater allergenic responses in humans can be constructed. This could lead to widespread concern, even in the absence of any evidence. In contrast, this new knowledge could actually reduce risks of allergens in the environment. Type-1 allergic reactions, such as hay fever and allergic asthma triggered by grass pollen allergens, are a global health problem that affects approximately 20% of the population in cool, temperate climates. Ryegrass is the dominant source of allergens because of its prodigious production of airborne pollen. Bhalla *et al.* (1999) found the major allergenic protein (named Lol p 5) of ryegrass pollen, which is present in the serum as immunoglobulin E (IgE) antibodies. Estimates are that nearly two-thirds of the IgE reactivity of ryegrass pollen is attributable to this protein. Bhalla *et al.* (1999) used the antisense technology to regulate the production of Lol p 5, engineering a new transgenic ryegrass plant with normal fertile pollen that was hypoallergenic. Hence, new technologies, while raising the need for new evidence, can produce new approaches (questions) that can mitigate risks and liabilities.

Finally, PMPs represent the ultimate state of scientific uncertainty, as neither the appropriate questions nor any relevant evidence is known. Chapter 10 examines this case further. The scale and scope of questions and answers in the fourth, last quadrant remain to be determined.

Implications for Policy and Liability

Transformative technologies, including plant biotechnology, are not part of 'normal science'. Therefore, activities in this sphere and its actors, scientists, regulators and citizens disagree on the questions and on how science is approaching the answers. Very few people accept that the K-K situation is secured, and many worry that we are really in the UK-UK situation of not knowing what we do not know about the new science and its applications. The effort to define the unknowns and to search for new knowns has triggered a wide debate about how science can and should support governance systems.

The four-quadrant model of knowns and unknowns applied against safety concerns related to pollens and plant biotechnology illustrates a number of points. First, if cause and effect are known and part of 'normal science', assessing, managing and communicating about risks is relatively straightforward. Where one of either the cause or effect is unknown, liability issues become self evident and governments are placed in the unenviable position of trying to find the best response, but they are often forced to react in some precautionary way. Where both cause and effect are unknown, blissful ignorance rules and governments often choose not to act.

Due to the large numbers of 'unknowns' within paradigms and because the instruments for gaining insights about them are shifting rapidly and often inconsistently, the public is becoming highly selective in accepting or rejecting questions and answers from the scientific community. Certain commodities derived from engineered microorganisms or health care products derived from unregulated parts of the industry go uncontested while GM foods are often rejected. As a result, the debate over GM foods has divided both the scientific and policy communities as much as the general public. Safety is no longer simply a scientific concept backed by scientific answers, it now has a major social dimension.

Policy makers therefore face an interpretative dilemma—what are the appropriate questions and answers and who should interpret and determine the acceptable risk level? Governments often act more from cultural and societal perspectives than solely from a scientific base. Governments have used the break in 'normal science' to adopt inconsistent and often conflicting policy positions (illogically ranking K-UK, UK-K and UK-UK cases) and, at times, inconsistently using precautionary rules.

Incomplete science has already posed two governance challenges for biotechnology. The first resulted from the work of Dr. Losey on the effects of *Bt* maize pollen on monarch butterfly larvae. Losey's study (Losey, 1999) was published in *Nature* and subjected to incredible media hype. The media story was that monarch larvae were dying after consuming *Bt* maize pollen. The research was incomplete and contained numerous research errors. The results of Losey's research contributed in part to a strengthening of refuge requirements in the production of GM maize. Smyth and Phillips (2002) provide a summary of the costs from increased regulation due to the incomplete scientific data provided.

The second challenge from incomplete science comes from the controversial research conducted by Ewens and Pusztai (1999). Pusztai conducted a series of tests on rats using GM potatoes. These researchers found that the consumption of these potatoes impaired the immune system of the rats and announced his findings on national TV in Britain, rather than submitting his findings to a peer review journal. The media sensation that resulted from his preliminary research contributed to the UK developing mandatory labelling requirements for all GM products. (See Smyth and Phillips, 2002, for additional details on the regulatory cost of Pusztai's research.)

In short, there can be three levels of liabilities: first (A), those that are disconcerting; second (B), those that require application of precautionary principles and perhaps disapproval; and third (C), where a significantly high level of liability would decide prohibition of the further use of transformative technology. As shown in Table 3.3, in the particular cases corresponding to Table 3.2, the ISP will belong to (A), and the CSP or PIAR to (B), requiring consideration of precautionary principles or disapproval. As yet, the PMP situation representing the UK-UK case is non-existent but prohibition or level (C) is the conceivable designation.

Table 3.3: Transgene pollen and liabilities

	Liability	
	Known	Unknown
Known	(A)	(B)
Unknown	(B)	(C)

Source: Authors

A different, but equally significant, issue in regulation of liabilities is the policy discourse. As a part of science, genetic engineering promises to make important contributions to food design and production, creating an agricultural framework that provides both environmental sustainability and food security (Khachatourians, 2002). In getting there, however, there are a number of discordant science and policy elements that must be addressed. The development and application of biotechnology for food or pharmaceutical plants have precipitated a major debate within and beyond the scientific community. There are many views about how to measure and secure safety as well as about how liability can be recognised, measured, contained or remediated. Outside the context of food, the pollens derived from transgenic plants can have larger environmental consequences. Pollination of wild populations with transgenes through the spread of transgenic pollen is not only highly undesirable but could be disastrous. Consequences could include changed vegetative growth, pollen-induced allergies, decreased generation times for tree breeding, increased fruit rather than seed production, altered quality or quantity of seed production, and even reproductive sterility in some forest trees.

Conclusions

There are many questions that science alone cannot adequately address. Where do the biological facts meet social truths? Are social responses to advances in transformative technologies influenced by 'normal science', experimentation and observation or paradigm shifts? How can or should a scientist examine re-interpretation and conjectures? As pointed out by Sackett *et al.* (1996) it is not clear that the current approach of compilations or critical reviews of the scientific literature will provide authoritative summations. These questions go well beyond the scope of 'normal science' but are fundamental to the effective governance of liabilities.

Finally, while objectivity in performing scientific enquiry is a must, the plant biotechnology debate has starkly revealed that scientific experts do not transmit objective information to policy makers in a way that will have a helpful influence on the formulation of policy. Beyond the inherent problems of paradigm shifts (creating new knowns and unknowns), scientists do not articulate scientific data for politicians in a manner that they can grasp objectively and use in the policy debate. As a result, emotion and rhetoric are often more influential than objective data. In addition, research programmes that attempt to address public concerns often have the opposite effect to that promised.

During times of paradigm shifts, politicians and scientists participate in consensual and mutually aggrandising promises and predictions. Without knowledge about the problems and answers, risk assessment, risk management and risk communications will founder. Calls for further research reflect a bias about the perceived role of science in policy making. The prevailing view is that science is there to solve problems. Instead, one could argue, using Kuhn's framework, that science is perhaps more important in defining the 'unknowns' and seeking new 'knowns', thereby creating the foundation for sound governance.

Nevertheless, one must be realistic about what science can do. The appeal to 'science' will not necessarily resolve disputes. We might assume that science speaks a universal language of 'truth' but at times it does not and perhaps cannot. Indeed the international science community is often unable to agree on the acceptable level of tolerances, which tests should be used and how results should be interpreted. In the absence of clearly delineated science, where knowns and unknowns are openly acknowledged and debated, policy makers, regulators, and even courts, will face a difficult time identifying risks and adjudicating liabilities.

Chapter Four:

Consumer Responses to GM Foods

Introduction

Economists and regulators have been unable to predict with any confidence how consumers might respond to GM products in the marketplace. Molecular genetics has revolutionised the agri-food business, converting the industry from a largely supply-push, evolutionary system towards a demand-pull, proactively engineered system. Biotechnology innovations entered global agri-food markets in 1995 with the introduction of rBST and herbicide tolerant maize, cotton, canola and soybean varieties. Since then the rate of new product introductions has risen sharply, with more than 17 crops already transformed by more than 50 traits.

A considerable number of consumers are or may be consuming these products. Maize, soybeans and canola are extensively traded throughout the world and their processed elements are essential ingredients incorporated into an estimated 70% of the processed foods on grocery store shelves. The inability of consumers to differentiate between products that either contain or do not contain GM ingredients has the potential to create significant commercial and socio-economic liabilities. Some segments of society in all countries have a preference to avoid consuming foods with GM ingredients and, if choice is not available for this group of consumers, then they may initiate either direct action (as through boycotts) or legal action against the food processing industry as a way to enforce their wishes. Should this group of consumers consume products containing GM ingredients, when they do not want to, they may also seek legal compensation for having been 'forced' to consume a product or products that they have no desire to consume.

Consumer attitudes and preferences regarding GM products vary widely both within and between key markets, reflecting the differing views about human health, environmental safety and product quality. People with different experiences and interests have widely different perspectives on these products. The 2002 Eurobarometer (Gaskell, *et al.*, 2003) found that while still more than half of respondents oppose GM foods, support among consumers in many European countries has risen in recent years. Support for GM foods in Spain,

Ireland and Finland was 70% or higher, while support in Greece, Italy and France was at 40% or less. Recent surveys in the US show that while opposition to GM foods there is lower, a large percentage of the public are still unaware of GM food products. In a recent study, Hallman *et al.* (2003) found that 43% of Americans had read/heard nothing or not much about GM foods.

This chapter examines the background, theory and practice of marketing and labelling GM foods, seeking to provide insights into how uncertain consumers can make and are making choices about the technology.

Background

The introduction of transgenetically-modified foods and food ingredients into the global agri-food system has raised public concerns about the long-term safety of these products for human health and the environment. It has been estimated that these foods (which have been cultivated on more than 58 million ha so far in more than 20 countries) and related food ingredients (e.g. proteins, oils, starches and sugars) are probably used in the production of almost 70% of the processed foods available. While regulators concur that there is no current scientific evidence that these modified foods involve any new or magnified risks, many civil society groups and some consumers are simply not convinced. In the absence of any definitive long-term studies, civil society groups and consumers have latched onto the idea that mandatory labelling for genetically modified (GM) foods will both empower consumers to select their own diet and will enhance long term monitoring and surveillance of GM foods.

These differing consumer attitudes are both a reflection of and reflected in national regulatory systems. In the EU, the general public does not appear to trust the institutions responsible for food safety. This decline in trust has developed over time due to previous food safety concerns but culminated with the BSE problem. At the time, in the early stages after the discovery of BSE, the British government and scientists told consumers that the beef was safe to eat and that there were no concerns regarding linkages to human illnesses—BSE was strictly an animal disease. As more information became available, the possibility of a linkage to human illnesses was posited and consumers were advised of the potential link. This change in policy only served to discredit the competence of food safety institutions in the eyes of consumers. Many Europeans have developed a risk adverse approach when faced with the choice of consuming products that are deemed safe but where there is minimal factual information.

In contrast, the consuming public in the US (and most other GM producing nations) exhibits higher levels of trust in the institutions responsible for public safety. This level of trust has been fostered by the actions that followed from previous failures. Following the Three-Mile Island nuclear accident, for example, regulations and standards were thoroughly reviewed and revised to ensure an even greater level of safety. Much of this was conducted in the public eye, which allowed those concerned to witness that the institutions were responding in a prudent manner to the concerns of the nation. The result of this process is that the American public tends to believe that their safety institutions have the ability to ensure that products in the marketplace are safe and that there is minimal risk to consumers when they purchase these products.

While the safety systems for evaluating GM foods may be effective, it is less certain that the governance structures in supply chains to manage and communicate information to consumers are adequate. The simple, yet hardly straightforward solution offered is to impose mandatory labels for GM elements in food. The rest of this chapter examines the theory underpinning consumer choices, reviews the current knowledge of how consumers actually perceive and value GM ingredients and critically examines the role of labelling in communicating with consumers.

Framework for Analysing Consumer Attitudes

The attributes of many of the new crop varieties create differing levels of risk and uncertainty in the market. Tirole (1988) provides one route to understanding how consumers might respond. He argues that there are three potential types of goods: search goods, where consumers can visually identify and select for attributes before consumption; experience goods, which require consumption to determine whether product attributes align with preferences; and credence goods, where the unaided consumer cannot know the full attributes of consuming a good, at least for some period after consumption. Trust usually is a key element in markets involving experience and credence goods.

In practice, a single product could embody attributes that fit all three types of goods. For example, if one is looking for a tomato, one could 'search' through the bins and find one that looks good, smells ripe and is apparently free of insects or disease. Once the consumer takes it home and eats it, they experience the quality of the fruit, judging it based on a variety of subjective factors, such as flavour and texture. Ultimately, the total utility derived from that tomato includes any longer-term benefits or costs of consuming the product, which as credence factors only become known some time after consumption. These could include some long-term benefits such as anti-oxidants, or some costs, such as food borne pathogens (e.g. *E. coli* or salmonella), which might become known within a few days, or toxic elements (e.g. carcinogens) that may have only a long-term cumulative effect on a person's health.

Many new crop varieties exhibit both experiential and credence elements, involving either input traits that entail some public concern (e.g. GM traits) or output traits that only have value if differentiated through more managed production systems. Consumer responses are inextricably linked to the production systems. The inability to search for the attributes necessitates more managed markets. The combination of significant potential for opportunistic activity (as some producers may wish to place low value crops into high value markets) and high asset specificity for many of the proprietary traits or for the output trait has created the incentive for a wide array of managed supply chains. Table 4.1 identifies where and why three different market structures may evolve in new crop markets. In the first instance, where there are no novel output traits or concerns related to the technology (e.g. no credence factors), spot markets are probably the most efficient market structure. When GM technologies are used, credence factors arise for some consumers, creating opportunities to differentiate and increase social welfare. Given the difficulty of searching for the identifiable traits, such differentiation will inevitably create specific assets and ample opportunities for cheating, necessitating a more integrated, managed supply chain. Finally, all output traits, whether conventionally or transgenically inserted, will generate significant challenges that will necessitate both public and private action. The two integrated chains will differ depending on the potential for real, measurable health and safety impacts. Where there are known or anticipated impacts, the state will inevitably reserve a role in managing the system while markets will probably be left to manage those systems where the concerns relate to preferences.

Table 4.1: Framework for optimal market structure for new crop varieties

	Non-GM variety	GM variety
Input trait	Search good; little potential for opportunism; low asset specificity; spot markets adequate	Credence good; significant potential for opportunism; high asset specificity; IPPM required
Output trait	Experience and/or credence good; significant potential for opportunism; high asset specificity; regulatory and IPPM systems required	

Source: Authors

The search, experiential and credence attributes of most foods are assured through a combination of public, private and collective labelling regimes (Table 4.2). In the production system, the public sector has tended ·to establish the general environment for private actors to effect transactions. The relevant food and drugs acts in various countries sets the rules for human consumption. At the retail level, consumer labelling laws have operated to establish consistency of standards around labels. Meanwhile, the private sector has established common-property or private mechanisms to match the transactional elements to the attributes. Companies employ trademarks, brands and warranties to assure customers of the value of their product. Experience has shown, however, that the costs of developing private standards are high wherever there are only competitive firms; for many agri-food products efficiencies have been gained by working through associations (e.g. Canola Council of Canada story, see Gray *et al.*, forthcoming).

In essence, commercial product standards can only really be understood in the context of all mechanisms used to manage markets (Table 4.2). At one extreme, governments or agents for governments set regulations to achieve public goals, such as health and safety or environmental objectives. At the other extreme, private companies develop brands and provide private warranties to assure consumers of the quality of their products. The consistent achievement of a high quality standard over a long period breeds a perception of quality, critical in the competition of knowledge-based innovative products (based on integrity and trust). Brand loyalty, a direct result of customer satisfaction, becomes a source of several long-term competitive advantages, which guarantee sufficient future demand. In particular, brand loyalty: is the basis for line extensions, a mechanism for enhancing revenue and reducing risk by capitalising on the perceived quality of the original product to facilitate market acceptance (Murphy 1990; Aaker, 1991; and Lane and Jacobson, 1995); increases the amount a customer will pay for a product in comparison with other comparable products, allowing for greater sales revenue through premium pricing; and contributes to responsiveness towards marketing efforts (Keller, 1993).

Table 4.2: Regulation, standards and private brands in labelling systems

Public Action	Collective Action			Private Action
Regulations for public good purposes	Regulations based on standards	Commercial and private standards	Private brands as standards	Private brands and warranties
Driver:	**Driver:**			**Driver:**
Public good market failures	Common pool goods requiring voice; collective rather than firm or regulatory based			Private, firm-based profit maximisation

Source: Authors

In some cases the regulations, brands and warranties become generalised and become the commercial standard for a commodity or product group, but most commercial standards these days evolve from collective action among producers with support by government. In this in-between area, there is a role to use standards in lieu of regulations to more effectively achieve the regulatory outcome. Standards can also help to facilitate trade based on product quality attributes. It is important to keep the drivers for standards separate, as the first set of standards (in lieu of regulations) ultimately are driven by public-good concerns while the second set are explicitly private goods.

Willingness to Pay for GM Foods

Given this theoretical approach, the challenge is to determine how much consumers would be willing to pay to have increased product or labelling information. The difficulty is that, wherever possible, most people want that information at no cost. Frequently, critics of biotechnology make the statement that 'labelling for GM is costless, all that is required is to put a label on the products'. Obviously nothing is costless, but the challenge is to determine what costs will be accepted by consumers. In short, consumer studies can go a long way to determining the aggregate value individuals place on this new information, and allow us to determine what would be the optimal amount of information that should be provided.

The recent literature in this field suggests that consumer willingness to pay for products labelled as GM can vary widely (Table 4.3). Rousu *et al.* (2002) examine whether there is a US consumer preference for products labelled with a 1% tolerance level versus a 5% tolerance level. This study found that consumption of products labelled as GM dropped between 7% and 13% regardless of whether the tolerance level was 1% or 5%. In fact, the authors suggest that there is no statistical support for US consumers having a preference for a 1% tolerance level over a 5% tolerance level. The conclusions reached from this research suggest that if the US (and probably other markets) were to adopt a tolerance level for the labelling of GM food products, then the socially acceptable threshold would be 5%.

Table 4.3: Comparison of survey and research results

Author	Date	Countries	Methodology	Results
Rousu *et al.*	2002	USA	Experimental auction	5% tolerance level for GM content is socially optimal
Moon and Balasubramanian	2002	UK and USA	Consumer survey data	UK – 77% prefer non-GM; US – 44% prefer non-GM
Kaye-Blake *et al.*	2002	New Zealand	Consumer survey data	Value of willingness to pay is NZ$285M and cost of GM labelling is NZ$42M
Grimsrud *et al.*	2002	Norway	Consumer survey data	Discount GM bread by 49% and GM salmon by 55%
McCluskey *et al.*	2001	Japan	Consumer survey data	Discount GM noodles by 60% and GM tofu by 62%

Source: Authors

Moon and Balasubramanian (2002) examine willingness to pay in the US and UK by offering consumers the choice between two identically-priced boxes of breakfast cereal, one box is labelled GM and the other is labelled non-GM. When asked which cereal they would choose if priced the same, 71% of UK respondents chose non-GM, while 2% chose the GM cereal and 23% were indifferent. Correspondingly in the US, 44% chose the non-GM product, 6% preferred the GM product and 22% had no preference. However, willingness to consume non-GM cereal dropped considerably when a premium for non-GM was introduced. Support for non-GM dropped to 56% in the UK and to 37% in the US when a small premium for consuming non-GM cereal was required. The number of consumers not willing to pay a premium (22%) was constant in both countries.

Research on consumer willingness to pay in New Zealand (Kaye-Blake *et al.*, 2002) examines the strength of preference for labelling by asking whether consumers would pay 2%, 5% or 10% more for groceries to learn about GM content. The authors found that 23% of the population would pay 10% more, 27% would pay 5% more and 24% would pay 2% more for product information on GM content. The aggregate value of the willingness to pay was estimated to be NZ$285 million annually, while the Australia New Zealand Food Authority estimated the annual cost of labelling to be NZ$42 million. A previous analysis by Smyth and Phillips (2001) suggests that estimates of willingness to pay should be treated with some caution. In comparative Canada-US analyses, those surveyed tended to respond more positively and be willing to pay more if they were presented cost impacts in percentage terms rather than in terms of dollars and cents per week. Thus, the cost-benefit ratio may not be nearly as wide as the Kaye-Blake *et al.* (2002) study suggests.

A study of consumers in Norway by Grimsrud *et al.* (2002) found that substantial discounts were required to get Norwegian consumers to purchase GM bread and GM salmon. The authors found that to entice consumers to purchase GM bread, a price discount of 49.5% was required and a discount of 55.6% was required for GM salmon. One interesting result from this study is that 26.8% of consumers would purchase GM bread and 17.8% would purchase GM salmon with no discount. However, 61% said they would never purchase GM bread at any discount and 67% would never purchase GM salmon. A similar study done by McCluskey *et al.* (2001) examined consumer responses to GM foods in Japan. This study examined willingness to purchase GM tofu and GM noodles and found that without a discount only 4% would purchase GM tofu and 3% would purchase GM noodles. The study determined that a discount of 60% was required to generate purchase of GM noodles and a discount of 62% for GM tofu.

This brief survey of the economic literature on willingness to pay highlights that every market has consumers that perceive and will purchase GM foods as an equivalent to conventional food purchases. On the other end of the spectrum there is almost always a group of consumers that will never purchase GM food products, regardless of the price differences. Those who are in the middle between these two groups (in many countries the largest group of consumers) are of most interest to researchers. While willingness to pay analyses could provide some insights into those consumers wishes, the studies will need to become more focused (perhaps eliminating those who are strongly opposed or indifferent) before they can provide much advice on how much one might be willing to pay for more specific labelling for either GM or non-GM foods.

Willingness to Pay for GM Labelling

The issue of labelling, whether it be mandatory or voluntary, spans the full spectrum of opinions. Critics of biotechnology and environmental groups constantly proclaim that over 95% of consumers in surveys demand labelling for GM content, while the biotech industry responds that, unprompted, only 2% of citizens in surveys call for mandatory GM labelling. Obviously, both of these numbers are being used to promote identified agendas and the real demand for labels probably lies somewhere in the middle.

Huffman *et al.* (2002) attempt to determine whether mandatory or voluntary labelling produces the most efficient economic outcome. In an experimental bid auction involving products that had basic labels, voluntary labels and mandatory labels for cooking oil, chips and potatoes, the authors found that the participants discounted the GM labelled oil more than the other products. One interesting result from this study was that in the mandatory labelling experiment the plain labelled products were perceived as the non-GM product, yet in the voluntary labelling experiment the plain labelled product was perceived as the GM product. This study concluded that for the US market, voluntary labelling would be a more efficient policy than mandatory labelling.

A study by Wolf and Kari (2001) found that US attitudes to GM products over four time periods between 1999 and 2001 were relatively constant. The authors found that in each of the four surveys, 80%-92% of respondents felt that the US government should impose mandatory labelling for GM products. One interesting feature of this study regarding labels found that 80% of consumers read nutritional labels when purchasing a product for the first time and 60% read ingredient labels in the same situation. Einsiedel *et al.* (2000) suggests that the percentage of consumers reading labels drops dramatically for subsequent purchases (to as low as 20% by some estimates).

Additional research by Wolf *et al.* (2001) concluded that many consumers do not understand the term 'GM-free' when included as product label information. Using simulated test markets for salty snack foods and fresh packed vegetables, the authors found that of the eight characteristics available to inform the consumer about the product, the characteristic 'free of genetically modified ingredients' was the lowest rated. This experiment also revealed that consumers, who prior to the experiment indicated that GM-free ingredients were extremely or very desirable when making a purchase decision, when faced with actual purchase choices did not express a higher interest in the food products labelled GM-free.

Recent studies on consumer attitudes in Colombia and Italy (Wolf and Pachico, 2002; Wolf *et al.* 2002) show much lower levels of consumer awareness towards GM food products. Surveys in Columbia show only 13% of respondents had any familiarity with GM foods while 77% reported no awareness. This research points out that 40% of consumers routinely do not have enough food to feed their families and that low prices were the most important factor when making a food purchase decision. This fact is important as almost 75% of consumers surveyed expressed food safety concerns about GM food products, yet nearly two-thirds of these consumers were willing to buy GM food products. Consumer surveys in Italy show that 28% of consumers are familiar with GM foods and only 43% would be willing to purchase GM food products. Interestingly, 40% of Italian consumers responded that they planned to increase their purchase of organic food products within a 12-month period of the survey.

An Agricultural Biotechnology in Europe (2002) poll on purchasing GM food products found that 52% of French respondents would continue with usual food purchases if the products contained GM ingredients. Those that were opposed to purchasing GM products were then asked if they would purchase GM food products if there were environmental benefits, and 47% of these respondents said they would. The remaining 25% were unwilling to purchase GM products for any reason.

As part of the ongoing polling of Americans regarding GM foods, the International Food Information Council (IFIC) released their eighth poll of consumers since 1997 (IFIC, 2003). This poll found that while 72% responded that they had either read or heard about biotechnology, only 36% were aware that GM food products are presently available for sale. Over six years of polling, IFIC has found that the number of consumers willing to purchase GM foods that had been engineered to taste better or be fresher ranged from a high of 62% in February 1999 to a low of 51% in October 1999. A new question for the polls done since 2001 asked for any additional information that consumers would prefer to see on product labels and in the four subsequent polls, only 1-2% responded that they would like labels to provide information on genetic modification

Critics of GM food products present numbers that represent the other end of the spectrum. In early 2000, The Council of Canadians commissioned Environics to conduct a national poll on attitudes to GM foods. Respondents were first asked a series of questions about their views on GM foods (such as 'I worry about the safety of GM foods'), to which 75% expressed concerns. Respondents were then asked if GM foods should always be labelled, and 95% agreed they should. In September 2001, Greenpeace released results from a Decima poll they commissioned and found that when asked if consumers had the right to know if GM ingredients were being used, 95% responded in the positive. When asked about whether the labelling system should be mandatory or voluntary, 87% wanted labelling to be mandatory.

Support for GM crops is evident in Europe. Gaskell *et al.* (2003) shows support for the use of GM crops in Spain, Portugal, Ireland, Belgium, the UK, Finland, Germany and the Netherlands. Overall, support in Europe is almost evenly divided, with 48% in support and 50% opposed. A recent IPSOS-REID survey of Canadian attitudes shows some interesting results (Johnson, 2003). Polling of Canadians on the issue of biotechnology and agriculture since 1998 has shown that concern about GM crops peaked in 2000 at 23% and has since dropped to 13%. While concern about GM crops is dropping, those that negatively view the technology of GM crops is rising. In 1998, those indicating their support for GM crops as very negative or negative was 45%, this has risen to 58% in 2003. However, when asked if they would support GM crops if the use of GM crops lowered chemical use, 78% would support GM crops.

Additionally, when asked about the safety of food in Canada, 80% responded that food in Canada is safe.

All of the calls and information about labelling can be put in context when consumer shopping habits are examined. A recent study by The US Produce Marketing Association (2001) summarising shopping habits for 2001 found that the primary factors for consumers when making produce purchases are: expected taste (87%); appearance (83%); cleanliness (74%); degree of ripeness (70%); and nutritional values (57%). This research highlights the fact that when consumers are in a retail store faced with a purchase decision, concern over whether the product is GM, non-GM or GM-free may be trivial for the vast majority of consumers. As a result, labelling systems alone would appear to be a poor way to communicate with consumers and would not substantially enhance post-market monitoring and surveillance.

Labelling Around the World

Governments around the world have responded in a variety of ways in response to widely varying demands from local consumers. Encouragingly, there are some common elements in the resulting labelling rules. First, all countries using the technology currently assess products for their human, animal and environmental safety, and do not allow GM foods with known risks to enter the food system. Second, all of these countries also have mandatory labelling rules for GM foods that involve nutritional or compositional changes (i.e. added vitamins or changed fat contents) or that have introduced new allergens into the food. Third, all these countries require that any label, whether voluntary or mandated by law, must be truthful and verifiable.

The differences, and ultimately the disputes, come when dealing with GM foods where there are no identifiable risks, or where there have been no changes in nutrition, composition or allergenicity. Labelling systems tend to reflect the domestic situation—while many consumers trust their food safety regulators, there are some consumer segments in most countries and majorities in some countries that have lost faith in their regulators or have religious/ethical concerns about the underlying technology. Those groups simply want labels to allow them to make their own consumer choices. Hence, the debate about labelling for GM elements without identifiable risks is mostly about what is the best way to provide consumers with information on the provenance of the foods that they eat.

Thus far about 35 countries have either adopted GM labelling laws or have indicated their plans for new GM food labels (Phillips and McNeill, 2002). All of the systems currently operating exempt highly processed foods like oils (canola, maize and soy), starches and sugars as they do not contain any detectable proteins. Also, none of the countries yet requires labels for products from animals fed with GM feeds (canola, maize and soy meal). There are two divergences in the systems. First, some countries—especially Canada, the US and Argentina, which produce GM foods—allow voluntary labelling, subject to either guidelines or general provisions related to honesty and verifiability, while a longer list of other countries—including the EU, China and Australia—have implemented mandatory labelling rules for all goods with detectable GM elements. Second, the EU is in the processes of adopting new laws to set more stringent requirements for labelling GM products, which may exacerbate the conflicts. Under the new regime, all products containing more than 0.9% genetically modified organisms (GMOs) will be labelled as GM products. Labelling will also be extended to animal feed, and to all products, from biscuits to frozen pizzas that contain highly refined ingredients. They also mandate full traceability of products, which will require farmers, manufacturers and distributors to collect and retain detailed information on the presence of GMOs in any product making its way through the commercial chain.

Thus, there are three emerging models of labelling—voluntary, mandatory for detectable traits and mandatory with full traceability—which would require different changes to the current food system and would impose different costs. Voluntary labels would appear to require the fewest changes in food systems and, hence, would result in the lowest overall costs. This type of system would deliver variety where either the incremental costs are minimal (such as by selecting ingredients without any current GM modifications) or where consumers are willing to pay any incremental costs. In the US, where voluntary labelling has operated for a number of years, there are an increasing number of food classes and individual products that are labelled as GM-free and, in a few instances, are labelled as containing GM ingredients. Mandatory labelling for detectable proteins, as currently adopted in the EU, Japan, China and Australia, can be relatively inexpensive—depending on the tolerances for co-mingling and the presence of GM free supplies—or can be costly, if they require development of dual production and marketing channels to segregate GM from other produce. So far no single country producing commercial quantities of GM food crops has adopted mandatory labelling. Countries have discovered that, while it is relatively simple to check produce at the border, it becomes very complex and costly to segregate and identity preserve within any existing supply chain. As long as some producing regions remain GM-free, the costs of this system should be manageable. Finally, mandatory labelling with full traceability, as proposed by the EU, will require a much larger group of products—e.g. highly processed ingredients and animal feeds—to be identity-preserved and traced, which will raise the costs further. If only the EU adopts this model, it may further impede trade between GM producing and exporting countries and the EU, but the aggregate economic impact could be manageable. The costs will accelerate exponentially if the EU exports its policies and standards, as it has been reported to be doing with Pakistan and other trading partners. If they succeed, mandatory labelling with traceability could impose a large one-time cost (some estimate 3% - 6% of annual food sales) and a significant sustained increase in the average consumer's food costs (some estimate up to 3% per annum).

The net result of these diverging labelling systems has been a reduction of trade between GM producers and exporters and those imposing stringent mandatory labelling, especially the EU, and a diversion of trade to other markets. This has tended to impose the costs of labelling not on the consumers seeking to know more about their food, but on the producers of those products.

While this might sound reasonable, it is just the opposite of how we deal with organic, 'Green', kosher and halal foods, where it is the consumer that pays. Given that these enhanced labelling systems for GM foods without identifiable risks are all designed to enhance consumer choice and not for public health and safety, then one might rightly ask the question of how to enable choice to those who want it at the lowest cost.

In spite of all the testing for GM ingredients, liabilities are still occurring in the food system. Continued testing for the presence of StarLink™ is still detecting the presence of this maize variety three years after it was supposed to have been contained and removed from the seed system. Clearly, when a liability occurs in the food system, the management of that liability can have long lasting impacts.

The power of consumers in the supply chain is only going to continue to increase and this new voice in the delivery of food products has to be dealt with in a more efficient manner. If clear choices are not available or if food products continue to show traces of GM ingredients when they are not supposed to have these ingredients, this consumer power could easily turn to liability lawsuits. Large class action lawsuits, such as those related to silicone breast implants, could conceivably become more common in the agricultural and food industries, as the consumers that desire to avoid these products seek more effective methods to ensure that they truly do have choice when purchasing food products. The potential cost of litigation and settlements may be enough of a deterrent to firms in the food processing industry to become increasingly diligent in ensuring that the sources of their raw ingredients meet the standards they have established for producing non-GM food products. If the food processing industry does not seek to impose these requirements on the rest of the supply chain, they may bear the brunt of the legal action brought against the food industry.

So far, discussion and debate have not brought convergence. There are two logical options. First, the GM food exporters could launch a trade dispute at the World Trade Organisation (WTO) against the evolving EU labelling policy, on the grounds that it is not based in science, there are no effective means to enforce the measure, it discriminates against imported products based on methods of productions rather than on products' characteristics and risks, and, ultimately, that it is more restrictive than necessary to fulfil a legitimate objective. A similarly constructed argument underlies the US led challenge of the EU moratorium. If a dispute follows, the process could take up to two years, and even then, if the EU lost it could still refuse to change its policy. Retaliation, which is the sanction for non-compliance with WTO Panels, only leads to increased trade barriers while not helping the injured export industry. Second, governments, industry and consumers could push harder and faster to develop standards for the labelling of GM foods. Currently, the national rules adopted diverge significantly, applying to different ranges of products and allowing different definitions of purity. One reason is that the rules imposed in almost every country so far have been arbitrarily established by governments, and have not involved consumers, citizens, industry or scientists in the identification of the thresholds and application. Canada is the only country that has attempted to develop a generally agreed upon standard. The Canadian General Standards Board in collaboration with the Canadian Council of Grocery Distributors, and more than 50 other interested groups has developed just such as standard, which should be operational in 2004 or 2005.

Conclusion

While mandatory GM labelling—in absence of or in place of more effective regulation—does not appear to be economically justifiable in very many (if any) markets, some alternative is needed to provide consumers with the information they are demanding.

As discussed, consumers need help in sorting through their concerns and to match their preferences with their consumption. Part of the solution is effective regulation (discussed in Chapter 5) part of the solution will be more comprehensive product differentiation (discussed in Chapters 8 and 9) and part will be more appropriate communication through product labels. As with most features of regulating biotechnology, to be effective, labels will need to fully and clearly identify public good type information (e.g. ingredients, health and safety information), private proprietary elements (through trade-names and warranties) and collectively developed and defended standards.

While international organisations (e.g. Codex) have spent considerable time and energy developing international voluntary standards for the labelling of GM food products, this initiative is still not providing consumers with the food choices they desire. Accurate product labelling, and, therefore, accurate product differentiation will continue to be a pressing issue for the agriculture and food industries. The potential for labelling liabilities is immense and both of these industries are being pressed to work more harmoniously to provide better consumer food choices in the near future (see Chapter 9 for a further discussion).

Part III:

Current Prescriptions

Chapter Five:

Regulating Transformative Technologies

Introduction

The introduction of a new technology or the application of an old technology in a new area is always likely to raise questions of who should decide whether the technology involves new or greater risks and how those risks should be assessed and managed. Biotechnology, a new and transformative technology with the potential to affect as much as 40% of the global gross domestic product (GDP), has such potential scale and scope of application that it has generated significant effort to ensure that the right decisions are made.

The advent of biotechnology in the global agri-food system has challenged our conventional conception of science, the social choices underlying the current risk analysis system and, ultimately, our confidence in the black-box nature of scientifically-based risk analysis managed by our regulators. Advanced breeding and biotechnology applications are making it possible to 'engineer' new differentiable traits in crops and animals and, at the same time, to test for and detect value-enhancing or value-reducing traits. Meanwhile, consumers and, by extension, processors, are becoming more demanding about the quality and provenance of their food. These new pressures have effectively destabilised the post-war consensus that underpins the global regulatory system for food, precipitating changing roles for governments and industry and new roles for other actors. The global agri-food system is currently grappling with finding a new consensus on how to manage new risks and new concerns related to these developments.

Ultimately, any risk analysis system ought to have a goal of unerringly rejecting unsafe technologies and products while allowing safe ones to enter the market. As with any human system, there is potential for error, especially when considering a new class of technology or product, where there are many uncertainties. While the system is and should be designed to avoid making type-1 errors—that is, erroneously approving something that is not safe—it has to be mindful of the trap of making type-2 errors—that is, rejecting safe products and activities (Table 5.1). Traditionally we have sought to avoid type-1 errors because we can usually tally up the cost of those errors in terms of lost lives, damaged ecosystems or other harms. Type-2 errors have often gone unnoticed as they represent forgone opportunities, where many of the key potential winners (e.g. new firms, employees and consumers) are not mobilised to present their case. Increasingly, though, there is potential that type-2 errors may become quantifiable.

Table 5.1: The typology of errors

Decision	If product is safe	If product is unsafe
Accept product as safe	Correct	Type-1 error
Reject product as unsafe	Type-2 error	Correct

Source: Authors

The litigious actions of firms who have felt wronged by government policy in the US and Canada, among other countries, have increasingly taken advantage of the domestic mechanisms of administrative justice and the international agreements detailing rights of investors (e.g. investment chapter of the NAFTA[1]), presenting evidence that challenges government decisions and quantifies the impacts of those decisions on their operations.

The conventional wisdom is that risk analysis is the natural domain of sovereign national governments, which independently analyse the impacts of any new technology and unilaterally decide, based on national criteria and capacity. In this context, industry, farmers, citizens and consumers all looked to their domestic governments to protect their interests. As international trade expanded following the Bretton Woods Agreement and successive rounds of the General Agreement on Tariffs and Trade, there was increasing pressure for risk analysis to be conducted consistently across connected markets. This internationalisation of regulation culminated with the World Trade Organisation Agreement in 1995, which has consolidated the scientific risk analysis capacity of a number of international organisations—i.e. International Plant Protection Convention (IPPC), International Office of Epizootics (OIE), Codex Alimentarius Commission (Codex), Food and Agriculture Organisation (FAO) and the World Health Organisation (WHO)—into the evolving liberal international trade regime. Earlier work (Phillips and Wolfe, 2001) argued that citizens, through local civil action groups, were increasingly pressuring governments to respond to this globalisation of risk analysis by renationalising parts or all of the analysis of new products. While some renationalisation has occurred, there is far more underway than a two-way tug-of-war. Practically, we

---

[1] E.g. Ethyl Corp. launched a chapter 11 (investment) dispute against the Canadian government using the North American Free Trade Agreement after Canada banned methylcyclopentadienyl manganese tricarbonyl (MMT) as an additive in motor fuel. Ethyl argued that Canada's decision to ban the additive without undertaking a science-based risk assessment had expropriated their investment (a monopoly position in the market) without compensation. Before the case was adjudicated Canada reversed its policy. No company has successfully prosecuted such a case, but the threat is real enough that the US government has raised the possibility of removing that provision in further rounds of negotiation.

have an array of powerful local (often parochial) institutions, in industry, government and various civil authorities vying with an equally powerful group of multinational enterprises, intergovernmental fora and global civil authorities to define or redefine the institutions that will make decisions about the ownership and use of new technologies.

The rest of this chapter provides an overview of the relationship between local and global institutions seeking to define and control the use of new technologies. It offers a brief review of the scale and scope of the new transgenic technologies that have been introduced into developed and developing country markets, examines various theoretical approaches to understanding the institutional challenge of risk analysis, reviews an array of examples of the local-global tensions between those institutions with power in the risk analysis system and offers some concluding comments and observations on implications of this struggle for the management of liabilities resulting from new technologies.

Sources and Uses of Crop Biotechnology

The international seed market has been the focal point for most crop biotechnology research in recent years. Researchers are attempting either to capture a larger share of the existing market (either by changing attributes or locking-in buyers through new, complementary packages of technology) or to expand the market by adding new traits. As one might anticipate, the focus varies depending on the investor. Only about 4% of the research and development (R&D) expenditures on crop biotechnology in 2001 were targeted at developing countries, even though they make up 35% of the global seed market (Table 5.2). Most of that research is undertaken by public agencies. In contrast, about 96% of the research (more than 70% financed by private companies) is targeted on developed country markets.

That research effort has resulted in 128 new trait crops, ranging from large acreage food, feed and industrial crops (e.g. canola, maize, soybeans and cotton) to small acreage vegetables, fruits, spices and ornamentals (e.g. potato, tomato, chicory and carnation) (Table 5.3). As one would expect from the location of the markets and the location of the research, most of these varieties are commercialised in developed countries (members of the OECD or the Cairns Group of 17 agricultural exporting nations).

Getting approval for release of a new trait crop does not necessarily lead to successful commercialisation. For example, while 11 countries have approved release of some crops (e.g. cotton), not all those countries are actually producing GM crops. Technical approval to release in many cases does not provide adequate justification to release. In some cases, there are ancillary laws that impose new liabilities on producers—such as mandatory labelling provisions in the EU—while in other markets the potential for unmanaged exploitation and the attendant risks of farmer saved seed, co-mingling and outcrossing create latent liabilities that impede commercialisation.

Table 5.2: Estimated global seed market and R&D expenditures on crop biotechnology, 2001

Size of seed market	$ millions	% of total seed market
Total	24,297	100%
OECD member states	15,790	65%
Cairns Group member states	3,375	14%
- of which OECD members	920	4%
India	970	4%
China	2,500	10%
Rest of World	2,582	11%

R&D outlays on crop biotechnology	$ millions	% of total R&D
World Total	4,400	100%
Industrial Countries	4,220	96%
- of which private	3,100	70%
- of which public	1,120	30%
Developing Countries	180	4%

Source: Seed market data from G. Traxler, *et al.*, 2003, drawn from FIS/ASA data sources; R&D data from James, 2002

One result of this uneven review, commercialisation and adoption is that a small handful of countries command the lion's share of the market. Over the first eight years of commercial cultivation (1995-02), the 17 approved GM crops were planted on more than 235 million hectares. Four main crops—soybeans, maize, cotton and canola—made up 99% of the total acreage planted to GM crops over the period. James (2002) estimates that the market value of GM seeds has risen from US$1 million in 1995 to US$3.8 billion in 2001 (if the average technology fee per hectare was US$25, then the total value of 54Mha in 2002 would be only about US$1.3 billion). Production is also geographically concentrated. While 20 countries have approved or produced one or more transgenic crop, the USA, Argentina, Canada and China accounted for 99% of the acreage over the first eight years of production (James, 2002). Given that most of these GM products are commodities subject to significant blending and shipping, it is often hard to determine the exact originating country of all parts of any shipment. As a result, any GM production that enters the international market could potentially be imported by any of the countries importing that

Table 5.3: Production and trade in genetically modified food crops

Commodity	Total world production 2002 # Countries	Volume (Mt)	Production (countries producing approved GM varieties) 2002 Countries	% total	Exports 2001 % total world exports from GM producers	Imports 2001 # countries importing commodity
Canola	53	33	Canada, USA	13%	47%	70
Maize	160	603	Argentina, Canada, France, Portugal, Spain, South Africa, USA	46%	84%	179
Melon	13	1	USA	0%	<1%	9
Papaya	51	6	USA	<1%	4%	75
Potato	154	307	Canada, Ukraine, USA	13%	9%	188
Rice	114	576	USA	2%	10%	193
Soybeans	84	180	Argentina, Canada, Romania, USA	59%	65%	126
Squash	83	15	USA	<1%	<1%	32
Sugar Beet	52	247	USA	10%	3%	34
Tobacco	129	6	China	38%	5%	193
Tomato	162	109	5*	37%	54%	148

*Note: *There are inadequate records to determine whether GM tomatoes are being grown in four other countries.*

Source: Adapted from Phillips (2002) and updated from FAO Statistic Agriculture Database, www.fao.org.

product. As noted in Table 5.3 up to 193 countries import some quantities of these 17 crops that have been modified. As a result, domestic and international trade rules, and ultimately private companies, are being forced to adapt to consumer and citizen concerns.

Approaches to Risk Analysis

Finding common ground for risk analysis is a complex process. This is at root an issue of institutional analysis. There is a wide range of approaches one might take. While scientists can report on what fits within Kuhn's paradigm of normal science and in the various regions of emerging or speculative science (e.g. where either theory or evidence is lacking, as discussed in Phillips, 2002), they cannot objectively define the appropriate level of risk. That is fundamentally a choice arrived at by social institutions—not by scientists operating independently.

Generally, there are three functions of big G and small g governance— legislation/negotiation, execution/management and adjudication/enforcement. While political scientists claim this area of study as theirs, in practice, economists and sociologists also have much to contribute. We posit that the institutional challenge of managing liabilities is actually the function of three interconnected domains: discrete local, regional, national and international state-based structures; local and global liberal economic markets; and an array of communal associations, clubs or social action groups that span the range from local to global. This section examines how one might define the scope, mandates, drivers and tools of each domain and the interconnections between their local roots and global reach.

Three convergent social studies are useful for highlighting the different aspects of human institutions. In the first instance, it is important to note that institutions encompass a wide set of rules, both formal (e.g. statutes) and informal (e.g. norms), which constrain relationships among individuals or groups (North, 1991). Institutions are not simply nominal rules—they are defined by those effective rules which can be enforced (Eggertsson, 1995). Institutions can be established, enforced and policed by an external authority or they can be voluntarily accepted, but the key is that they are predictable, stable, and applicable in repeated situations. Political scientists, such as Susan Strange (1988), talk about these norms as being formal or informal bargains, economists talk about them as market rules and sociologists reference them as social norms. Regardless of the framing device, it is important to keep in mind that institutions are not found in nature—they are human constructs.

Strange (1988) perhaps best articulated the international political economy (IPE) framework for analysing the challenge of managing the marketplace. As with most political scientists, her focus is on the sources and uses of 'power', i.e. the capacity to determine outcomes. In her book, *States and Markets*, she examined the dichotomy between states—which have the power to coerce directly with hard-power or through structural soft-power—and markets—which at root have a self-governing liberal economic framework but tend to concentrate to exploit market power—and how they engage in legislating or negotiating the production of goods or services, manage the resulting 'bargains' and adjudicate or enforce any disputes. In essence, her focus is on how various structural factors—i.e. her security, production, finance and knowledge systems—apportion latent power and how these actors enter into bargains to mobilise that power. In later chapters of that book, and in some of her later work, she suggests that the concept of states could be enlarged to 'authorities', which include both all the traditional state actors and a new set of civil actors with soft-power to establish civic norms. Perhaps a better way to look at it is to see this as a mix of three different types of actors—states, markets and non-state authorities—that strike bargains to mobilise their latent power. Each of these can create bargains with others in their domain or across domains in order to do so.

Meanwhile, sociologists have engaged in the analysis of the post-industrial, post-modern, knowledge-based world through the triple helix framing device (Etzkowitz and Leydesdorff, 1995). They see the world through social relationships. Generally, they believe that while explicit power is important, in a post-modern, liberal democracy, it is only effective if it conforms to the underlying social structure of the society. Giddens (1994) and others argue that what they call 'mode-2' knowledge in a post-industrial and post-modern world is created, owned, controlled and used through an array of formal and informal relationships between three main actors—governments, industry and universities. Each has specific roles, interests and capacities to facilitate knowledge-based growth. Just as IPE is groping towards a unified framework for analysis, sociologists have been examining their approach and considering refinements to explain recent developments. The main challenge has been to explain the rise of non-governmental, civil or social action groups, some at the local level and some at the international level. Mehta (2003) has argued that these groups represent a fourth strand in the helix, or a fourth helix. Just as for IPE, one might consider the possibility that these four might distil down to three functional types: governments that engage in political discourse with citizens; markets that engage in economic exchange; and a variety of collegial or social groups—including universities and NGOs—that engage in social discourse among sub-groups in society.

Last but not least, economists have characterised the problem as one of managing commercial transactions related to three types of goods—public goods, private goods and club or association goods (Gray *et al.*, forthcoming). Economists posit that in a competitive marketplace made up of many informed buyers and sellers, arms-length market exchange is an institution that effectively governs the production and consumption of goods and services. The prices generated by the selfish actions of a large number of actors in a market create Adam Smith's 'invisible hand,' whereby the marginal cost of providing a good is matched to the marginal value of that good to consumers. In a great many instances in the market place, a simple exchange of goods and services at an agreed upon price is a low-cost transaction that provides the correct incentives for buyer and sellers. In some cases, however, the costs of searching, negotiating and enforcing transactions through the marketplace are too large, leading to new institutional arrangements. As mentioned in Chapter 1, Picciotto (1995) posits that particular institutions tend to be best suited to govern particular types of transactions. He classifies institutions into three general types—the government sector; the private sector; and the participatory sector. Each sector represents different individuals, involves different incentives and is effective in producing goods or attributes with specific characteristics. Each of the theoretical approaches provides some insight into the array of potential institutions that might influence liabilities. Effectively, each offers an explanation of how and why different actors might lead in the legislation/negotiation, execution/management and adjudication/enforcement of new arrangements to manage the potential liabilities from the introduction of new products.

The Practical Regulation of Liability

The practical regulation of potential liabilities of new, transformative technologies is inevitably going to be complex, as it often involves highly uncertain impacts.

The safety evaluation systems operating in OECD countries are, for the most part, scientifically-based processes that combine the identification and characterisation of hazards with assessments of exposure to characterise risk. In essence, they purport to objectively assess the probabilistic outcomes of discrete adverse events, for the most part abstracting from issues related to risk management or risk communications. The practice is that different actors— governments, industry or civil society—establish a risk threshold for technologies, products or classes of products that would enable them to reject new products with unacceptable risks but would allow those with acceptable impacts to enter the market. This activity begins at the earliest stages of research and continues through to post-marketing liability management.

Even before any investment is made, individual companies, government granting agencies and civil action groups begin the process of debating and formulating opinions about the range of activities that would be acceptable in the environment, economy or society. Some of this effort is ad hoc, with individual project proponents making decisions on what to pursue or not to pursue based on their own estimation of potential liabilities or on their own value systems. Increasingly though, this process involves engaging civil society in a more formal review, either by governments through their granting council ethics boards or by industry through a range of scientific and ethics advisory boards or contracted advice from experts. Ultimately, many of the potentially most risky and uncertain outcomes are avoided at this stage.

The process of managing risk and avoiding potential liabilities continues mostly informally throughout the research and development phases of the majority of projects. Except in a few discrete cases, governments have left risk management in the R&D phase in the hands of the researchers themselves. Nevertheless, there is a wide range of actors involved in this. Individual companies have invested heavily in managing risks at this stage, as they would be exclusively criminally and civilly liable for any errors that cause harm. Meanwhile, professional associations (e.g. chemists and biosafety groups), standards organisations (e.g. the International Standards Organisation [ISO] and Good Laboratory Practices [GLP]) and various NGOs help to define standards or offer quality assurance systems to more effectively manage risks in R&D.

Once a product or technology is ready to be commercialised, more formal government-managed risk assessment commences. The Food and Agriculture Organisation define risk assessment as "a scientifically based process consisting of the following steps: (i) hazard identification, (ii) hazard characterisation, (iii) exposure assessment, and (iv) risk characterisation." The actual structures and procedural standards for scientifically-based risk assessment (SRA) were recently agreed and adopted as an international standard of the Codex Alimentarius Commission of the FAO/WHO (Codex, 2003).

Different portions of this analysis are defined, managed and adjudicated by experts and non-experts operating under a wide range of institutional arrangements. In some cases an expert may use their own expertise and values independently to characterise the hazard and determine whether the exposure is acceptable, but in most cases they rely on a broader base of expertise that resides either in their professional group (sometimes codified and promulgated but more often simply tacitly acknowledged and available to those skilled in the art) or to outside procedural standards or tolerance levels (adopted either by government, industries or broader civil groups). Increasingly, non-experts are defining, managing and adjudicating the risks.

While domestic regulatory systems currently provide the foundation for regulating safety, this necessarily involves a wide array of government agencies, producers, processors, wholesalers, retailers and, ultimately, consumers. The OECD (2000) recently completed a major review of national food safety systems and activities that highlights the similarities and differences of the various domestic systems. At root, all of the countries in the report are somewhat similar in that they base their food safety decisions on a scientifically-based risk analysis approach, which is understandable in that these systems have evolved over the past 50 years in tandem with the development of international agreements on standards and procedures, embedded in the IPPC, OIE, Codex and various other technical agencies of the United Nations (UN) or the Bretton Woods system. The differences lie in how they implement their risk assessment, risk management and risk communications.

Each of the member states in the OECD generally has some relatively well defined set of criteria for evaluating risk and has a system for undertaking that review. All countries indicate that they use scientific risk analysis as part of the assessment. The US, at one extreme, emphasised that their decisions are always based on objective scientific evidence and that their decisions are subject to procedural appeal through the courts. Other countries indicated that an array of more subjective, socio-economic criteria could be considered in the review of risks. In many cases it would appear that these other criteria are inserted at a more political stage of review (e.g. until recently through the Article 21 Committee in the EU). The evidence available suggests that for the most part the scientific judgements of risk by regulators in different countries tend to be very similar, whereas the political judgements have potential to vary widely. As for the process for review, most countries have an array of agencies and legal authorities involved. In the past there was a tendency to divide the responsibilities among different authorities, with separate acts and regulators for each stage of the food chain or for each type of food product. Increasingly the trend is to develop more centralised authorities. The US has the Food and Drug Administration, Australia and New Zealand created the ANZ Food Authority in 1991, Canada created the Canadian Food Inspection Agency in 1997, the UK created the Food Safety Agency in April 2000, the EU has launched the

European Food Safety Agency and Japan created a unified Ministry of Health, Labour and Welfare. Meanwhile, most of those countries have reviewed or are in the process of reviewing and consolidating their legislation. More problematically, all of these regulatory systems depend critically on the involvement of an array of participants from the food system. As risks are often specific to the technologies and processes used, individual companies and sectoral associations are key suppliers of information that regulators can use to evaluate the safety of the food. Given constrained budgets and growing workloads, one challenge all governments face in assessing risk is that rapidly advancing science may be beyond regulators' assessment capacity. National regulators may not have adequate resources to fully assess the particular risks of products submitted for review, which could increase the likelihood of either type-1 or type-2 errors.

Risk management tends to be less centralised in most countries. Given the federalist nature of many of the states—especially Australia, Canada, EU and the US—and given the practicalities of monitoring a wide range of foods across large areas and often across international boundaries, most countries have adopted extensive networks of assessors, auditors and inspectors to maintain the safety of the food system. Most of the major exporting countries—e.g. Australia, Canada, New Zealand and the US—indicate that they are also aggressively moving towards more industry involvement in food safety, through industry-based standards systems such as Good Manufacturing Practices (GMP) or Hazard Analysis Critical Control Points (HACCP) systems.

Risk communication is less organised (or at least less well communicated!). Most OECD countries identified that they have systems for communicating with consumers about food safety risks, but most of the systems do not appear to be formal or comprehensive and all depend on the commercial media to get their message out. While the increased use of the Internet for posting information and decisions is helping to increase the transparency of the system, more could be done. The US goes the furthest in committing to communicate during the risk assessment process, with its requirement to post draft decisions in the Federal Register. Other assessment systems are more opaque until decisions are reached, and even then the decision documents are often not detailed enough to critically assess the basis for the decision (Doern and Reed, 2000, discusses this problem in Canada). Communications related to risks in the marketplace are equally challenging. Labelling is the primary formal vehicle for conveying information. While all countries require labelling for contents and for specific scientifically verifiable product risks, such as nutritional or compositional changes in a product or the presence of new or modified allergens, many of the countries do not even require labels for nutritional elements. Manufacturers voluntarily label a wide range of food features, such as nutrition, country of origin, production method (e.g. organic) and other socio-economic or ethical features (e.g. animal welfare, kosher, halal). Recently, Australia, the EU, Japan and South Korea, among other countries, have announced or adopted mandatory labelling for GM foods (see Phillips and McNeill, 2002, for details). Many regulators and food industry companies are trying a variety of other means of communicating with consumers, including point of sale information, but consumers still express confusion and uncertainty about the risks related to the food they eat (Einsiedel *et al.*, 2000).

The advancement of new technologies, specifically biotechnology, has forced many of the domestic regulatory systems to re-evaluate their risk analysis processes. As is obvious from the foregoing discussion, all OECD countries have some form of domestic food safety regulation that predates the arrival of biotechnology in the global production and trade system. Depending on the degree of concern in the domestic market, countries have either relied upon those systems, in conjunction with the regulatory oversight of the exporting country, or have built new or modified systems to handle their specific concerns about biotechnology. Two divergent regulatory approaches have evolved. The North American regulatory approach (generally used by Australia, Canada, Japan, Mexico, New Zealand and the US) is relatively pro-supply, relying on a legalistic interpretation of mostly scientific assessments, while the European model pursues a precautionary approach that reflects citizen and political concerns (Phillips and Khachatourians, 2001). Although both approaches delivered similar decisions in early years, they have diverged significantly since 1998, with the result that more than 36 GM food products have been approved in the US that are not approved in the EU.

As a result of the incomplete and often conflicting domestic regulatory systems in both exporting and importing countries, the regulation of GM foods has become an international issue.

A Practical Example of the Challenge of Liability Management

Perhaps the best way to illustrate the current challenge is to examine the global regulation of one specific technology. For illustrative purposes, we have chosen to examine the regulation of risks related to the introduction of transgenic, herbicide tolerant (HT) canola.

Each of the three transgenic, HT canola events have been first approved and commercialised in Canada, but the effort to create, commercialise, assess, manage and communicate the risks and liabilities has involved a wide range of actors from around the world. While maize, cotton and soybeans are, on a first inspection, products primarily of the US, their research, development, commercialisation and management have similarly engaged a range of state, market and non-state actors in multiple countries around the world.

The initial research for canola involved a global team (Phillips and Khachatourians, 2001). Between 1981 and 1996, approximately 6,900 scientists in approximately 1,500 organisations in 79 countries undertook some work on canola. Furthermore they cited 17,995 papers from 1,294 journals, produced by approximately 28,800 authors in 3,816 organisations in 107 countries as producing research that was relevant to their work. In addition, the canola researchers contributed to and drew upon the art and craft codified in more than 600 patents involving canola-related technologies and products. This knowledge was assembled in the late 1980s and early 1990s by a relatively small group of researchers—780 world-wide, 400 of which are in Canada—led by an elite group of about 40 star and emerging star scientists[2], 18 of which practise in Canada. This group of scientists were bound together by a web of overlapping and interlocking memberships in academic disciplines, a wide array of formal and informal collaborative research initiatives and involvement in a wide range of professional, commercial and public service ventures. Beyond prevailing social norms, these relationships are moderated by the general scientific method that underpins normal science. Normal science operates where there is a comprehensive paradigm (with both theory and methodology) that offers predictions that are borne out in the evidence (Kuhn, 1970). This science is almost exclusively the preserve of the global academic scientific community, being developed, stored and transmitted from universities and other peer-reviewed research centres. As such, the internal control systems in the global scientific community—namely the peer reviewed structure of appointment, tenure, promotion and publication—enforce rigour and conformity.

Meanwhile, these scientists work in institutions that offer some more overt and explicit command and control structures. Most institutions now involve ethics and safety review and intellectual property analysis as part of evaluating research proposals. In the case of transgenic, HT canola, research was undertaken for the most part by public researchers in Agriculture and Agri-Food Canada, the National Research Council and a variety of public universities, often in collaboration and partnership with the private companies that owned many of the core intellectual properties and financed a significant share of the research. By using public funds or facilities, they triggered provisions of Canada's Tri-Council ethics framework, Treasury Board investment and financial guidelines, provisions of Canada's Public Servant's Invention Act which specify ownership and management responsibilities for new technologies or products, and a variety of private standards for ethical or risk management (including conformity with various industrial standards which are embodied in the Good Laboratory Practices codes). As one might expect based on the multinational character of the firms involved (i.e., Monsanto, AgrEvo/Bayer, Cyanamid), many of the industrial standards were international.

By 1988, transformed plants with new transgenic traits were available and ready for testing. All new varieties of grain and oilseeds generally flow through the same system in Canada, with higher levels of oversight on those that involve novel traits. Whereas private or public breeders were responsible for managing any risks in their research programmes as long as the materials remained in isolated conditions (e.g. in laboratories or under glass), once the breeder has developed a cultivar that was stable and unique and was ready to have it examined for registration, the formal system took over.

The first step was to make an application to the regulator—at that time Agriculture and Agri-Food Canada (AAFC), now the CFIA—to undertake confined field trials to test the variety in the environment. The agency examined the traits involved and established guidelines for the trial, including isolation distances, weed control provisions and auditing. The trials were designed to

---

[2] Adapting from Zucker *et al.* (1998), stars are defined as those having published 20 or more peer-reviewed journal articles that were cited on average 5 times each; emerging stars had 15 articles cited on average 5 times or more.

provide the evidence to evaluate the environmental risks of the new cultivar and to assess its agronomic merit (e.g. yield, disease resistance, time to maturity, quality and other traits). This involved developing with the academic community and industry a regulatory directive on the biology of the species. The regulators also needed to see evidence on the characterisation of the transformation system, the nature of the carrier DNA, genetic material delivered to the plant, the components of the vector and a summary of all genetic components. In addition, the regulator required an array of data to assess the inheritance and stability of the genetic modification (e.g. Mendelian segregation) and a description of the novel traits (e.g. Southern analysis and qualitative ELISA analysis of the gene expression levels) (see www.inspection.gc.ca/english/plaveg/bio/subs/subexe.shtm for a detailed list of what this involves). In the case of canola, these characterisation processes were being developed simultaneously in the US, Canada, Japan and EU, with the companies and their public sector research partners providing data and methods, and the regulators working individually and collectively to structure the procedures to enable them to assess it. Since 1998 the Canadian Food Inspection Agency and the United States Department of Agriculture (USDA), Animal and Plant Health Inspection Service (APHIS) have also been studying and comparing the molecular genetic characterisation of transgenic plants in a search for ways to harmonise their regulatory review processes. Some agreement has already been achieved although no formal binding bilateral agreement has yet been concluded. The various member states of the OECD have also been contributing to the development of Consensus Documents, which are scientific background documents mutually recognised by member states that set out the biology of the crop plant, introduced trait, or gene product, and provide a common base to be used in regulatory assessment of an agricultural or food product derived through modern biotechnology. By October 2003, 28 consensus documents were published by the OECD (Document 7, 1997, OCDE/GD[97]63 is the relevant consensus document on the biology of canola). Meanwhile, the IPPC has addressed the international regulation of GM crops through several International Standards for Phytosanitary Measures (ISPMs).

Once confined field trials were authorised, they were undertaken following a strict set of guidelines and standards, which, while national in application, were drawn from international evidence of the appropriate risk management procedures and the latest international biosafety evidence. While the regulators were responsible for auditing and enforcing the rules on trials, these trials were usually managed directly by the research firm or by a contractor (in Canada, the various research farms operated by AAFC have managed many of the trials under contract with the companies). While there has been some effort to internationalise these rules through the IPPC, some differences remain. Between 1995 and 1998, 497 field trials on canola were conducted in 12 countries (almost 300 of those trials were undertaken in Canada; the bulk of the rest were in France, the UK, the US and Belgium).

By 1992, the companies had gathered enough data to demonstrate intergenerational stability, agronomic efficacy and commercial promise and began to develop their regulatory package of evidence to present to the regulators to assess the safety of the products. In Canada, this required extensive data on the toxicity of the novel gene products (e.g. a series of toxicity studies with humans, animals and non-target species). The product proponents also had to provide scientific studies on the nutritional aspects of the novel trait and plant for both humans and livestock and comparisons of the amino acid sequences of the novel trait to known allergen proteins. Finally, the proponents were required to provide a package of studies on the environmental impact of the novel traits on soil, weeds, wild relatives and non-target organisms. McHughen (2000) published a histogram of the volume of data required to satisfy regulators of the health and safety of transgenic crops (in his case a transgenic flax variety)—the pile of studies and reports exceeded 6 feet for the transgenic product, versus an average of about 30 pages for a conventionally-bred variety. Given the array of studies and evidence required, there is necessarily extensive reference to global science and, perhaps more fundamentally, to contracted research by academics and other scientists.

The results of the field trials, food, feed and environmental reviews were then examined by the appropriate regulators. In Canada, Health Canada undertook the food safety review while the environmental and animal health reviews were conducted by forerunner agencies of the CFIA (in the US, the Federal Drug Administration (FDA) and USDA/APHIS would take the lead). In each case they had enabling standards embedded in legislation or regulation which needed to be made specific for each product or technology. That process involved extensive negotiation between the regulator and the product proponent, supplemented with reference by the regulator to experts in other national regulatory systems and to those outside the regulatory system. As Mills (2002) points out, there are often legitimate disagreements that can unhinge the process. Her analysis of the differing rBST decisions in Canada and the US highlighted how differing roles of science—regulatory vs research science—and differing methodologies—statistical versus biological significance—can fundamentally alter risk assessment decisions. Specifically, her evidence shows how different contextual values can translate the same science into contradictory regulatory decisions.

Finally, the three new trait canolas were assessed by a committee of researchers operating under the authority of the Canadian Seeds Act—they analysed the candidate varieties against a 'check' variety—and then the committee authorised them for sale to farmers. Most other countries do not have this regulatory step. At that point a blended public-private quality control system took over. Most new varieties were multiplied either in contra-season locations (e.g. Chile, Australia, southern US states) or by registered seed growers in Canada. The Seeds Act identifies the Canadian Seed Growers Association (CSGA), the umbrella organisation of provincial growers associations, as the official seed pedigree agency in Canada responsible for certifying all new varieties and ensuring they meet the standards set in the Act. They establish the rules for producing foundation, registered and certified seeds and then undertake the audits and conformity measures of growers that are necessary to ensure the standards are achieved. Even so, most seed companies also established some supplementary rules and procedures for contract registered-seed growers to ensure that their varieties are properly multiplied. Alternatively, some companies multiplied the seed themselves on company farms.

Once the seed was sold to farmers, the risk management/quality chain became much more complex. Apart from any regulatory requirements (e.g. contract registration rules for a few varieties that specify isolation distances) and any private contractual obligations (e.g. crop rotation rules, agronomic advice and record keeping), farmers are for the most part allowed to use their own judgement to manage the quality of their crops. There are currently no formal, on-farm, quality assurance programmes for grains or oilseeds operating in Canada or elsewhere.

Once the grain is called forward into the grain handling system, the quality system again becomes more formal. While Canadian law, analogous to rules in other countries, places the onus on the developers, owners and users of any technology to monitor and report any post-marketing adverse effects, there is no formal auditing or research undertaken of their effectiveness. All grain entering the handling system in Canada is graded under the authority of the Canadian Grain Commission (CGC), which is recognised by the *Canada Grains Act* as the official organisation to set and enforce the grain grading system. (Other countries leave this step to the commercial trade). The CGC, after consulting with the domestic grain trade and international buyers, sets the standards for the grading of grain for domestic and export markets and licenses the grain elevators and shippers. Even so, the grain merchants, railways, export terminal operators and shippers co-ordinate the logistics services for the private merchants that move the oilseed to the export customer. Hence, their procedures and practices have a vital role to play in assuring the integrity of crop quality. Ultimately, private processing companies take over responsibility for risk and quality when they take possession of the grains and oilseeds and process them into food products. Apart from pro-forma regulatory standards for food safety and laws requiring honest disclosure of contents and weight on labels, the processing companies are left to establish and maintain their own quality standards. Often those standards are set and audited by retail chains.

The introduction of transgenic canola in Canada highlighted the degree of integration of regulators, industry and associations. In 1995 both Monsanto and AgrEvo received approval from the Canadian government to commercialise new transgenic, herbicide tolerant canola varieties. Although the US approved the products almost simultaneously, Japan and the EU, which jointly accounted for approximately 42% of the market for Canadian canola in 1994/5, had not yet approved the varieties. After representations by the Canola Council of Canada, an industry association involving producers, processors and traders, the two biotechnology companies worked to develop an identity preserved production and marketing system with the Council, Canadian regulators and the oilseed wholesalers and processors. Once the system was developed, Canadian regulators, the biotechnology developers and representatives of the Council visited regulators and importers in Japan and the EU to answer questions and assure the importers that no GM canola would enter their supply chains before the assessments were completed in those markets. By all accounts this system was effective, thereby enabling adoption of the technology to occur two years earlier than would otherwise have been possible. The accelerated adoption of this new technology is estimated to have generated net benefits of about C$100 million for innovators (as discussed in Chapter 8).

There are continuing efforts to knit together these disparate and diffuse sources of knowledge, expertise and capacity into a more effective system. There are both bottom-up and top down efforts underway. While many of the learned societies and national advisory groups (e.g. the Royal Society in the UK and the National Research Council and National Science Foundation in the US) are developing assays of the scope and nature of normal science related to GM foods and industry is developing its own set of collective institutions from below, national and international governmental organisations are trying to pull things together from above.

Recently the agrochemical and crop biotechnology industry in Canada has more proactively managed pre- and post-market risks through two CropLife comprehensive stewardship programmes—one for agro-chemicals and the other for biotechnology products. Both of these programmes are under the stewardship*first*™ umbrella. More than a continuously improving set of standards and codes, it is a guiding philosophy of proactive and evolving self-regulation. The stewardship*first*™ agri-chemical programme encompasses initiatives that manage the stewardship of crop protection products through their full lifecycle in order to protect both the population and the environment (Hepworth, 2001). The second stewardship programme area is the Biostewardship initiative, which when complete is expected to involve industry protocols for safety in labs, greenhouses, field trials, transportation and international movement during R&D, development of a curriculum and training programme for novel-trait confined-field-trial managers, standard bag labels for seed, segregation and traceback systems, distribution channel training, marketing codes, manufacturer and farmer training and certification to international standards (e.g. ISO, HACCP), and rules for disposal of unwanted seeds and empty bags. If this full system becomes operational, it will tighten the supply chain spanning between researchers and the end consumer, limiting the potential for liabilities to compound.

Meanwhile, the major countries developing and using biotechnology are engaged in an extensive exercise to establish some top-down harmonised systems. In addition to the efforts to establish acknowledged intergovernmental standards through the IPPC, OIE, Codex and OECD systems, there are extensive bilateral efforts. The Trans-Atlantic Economic Partnership (TEP) between the US and EU, for example, has undertaken talks in recent years to improve regulatory processes and scientific co-operation through: mutual recognition of testing and approval procedures; progressive realignment or adoption of the same standards, regulatory requirements and procedures; adopting internationally agreed standards; and dialogue between scientific and other expert advisers in standard setting bodies and regulatory agencies. The EU has similar trade liberalisation initiatives with Canada and Japan.

Many are unhappy with the continuing discontinuities in the international regulation of these products. The next chapter examines the issues, approaches and unresolved elements of an international consensus based on the World Trade Organisation and the Convention on Biological Diversity.

Conclusions

This new world generates three complications. First, the concept of sovereignty that came from the Treaty of Westphalia of 1648 has become increasingly irrelevant as the science and the markets have become globalised. It is almost impossible to talk about any country independently developing, regulating or producing any knowledge-based good or service. As illustrated above, every country, including the US, requires the know-how, codified knowledge or technical expertise resident in other countries, multinational firms or international scientific and regulatory bodies. Thus, sovereignty is diminished by the need to rely upon others, especially those that cannot be influenced by any nation state. While multinational firms and international regulatory bodies at one level are accountable or can be held to account by at least one nation state, the international scientific community is largely beyond the control of any government or group of governments working together.

Second, transparency, a common standard for public policy and a principle that is increasingly incorporated into international agreements (e.g. WTO Agreement and the Cartagena Protocol on Biosafety), is becoming difficult to achieve given the dense and complex web of actors and relationships involved in assessing risks and managing liabilities. Even when governments commit to maintaining full transparency, the sheer volume and complexity of the data, methods and procedures for evaluating new technologies are making it very difficult to figure out the sources and uses of evidence. Compounding this is the frequent cross-referencing of evidence between domains and the all-too-frequent imposition of secrecy provisions triggered by 'confidential business information'.

Third, the twin concepts of responsibility and accountability that underlie modern pluralist democracies are elusive, at best, in a period of eroding sovereignty, opaque management and diffuse authorities and actions. It is becoming next to impossible to assign blame to anyone for a regulatory or marketing failure. The responsibility for any discrete action is often shared among many actors, including different governments, firms, scientific bodies and international organisations.

Chapter Six:

International Governance of Liabilities

Introduction

The evolution of multilateral regulatory regimes is a slow process that is based on consensus-building among a large number of participants. It suffers from the logistical difficulties of dealing with the need to allow input from each of the participants in a timely fashion, to allow for the interplay of argument, persuasion and the blunt political 'horse trading' that constitutes the negotiation process and to accommodate the requirement that whatever compromise is reached in the international forum must be accepted by the domestic political bodies of the participating countries. The World Trade Organisation, for example, has in excess of 140 member states. Add to this the possibilities for 'hold-up' that decision-making by consensus allows and the impossibility of excluding the influence of the wider international political climate on any deliberations, and it is sometimes surprising that anything can be accomplished. In recent years, there has also been a rising concern regarding the capacity of developing countries to participate effectively in international negotiations both because their negotiators lack the technical support to deal with complex issues and their governments are unwilling to commit sufficient resources to ensure full participation in the multitude of multilateral decision-making forums. The need to build capacity has further slowed the process of international consensus building.

In short, the inherently slow pace of decision-making means that international organisations function best in relatively stable international environments. They are least effective during periods of rapid change. By definition, a transformative technological change such as the advent of biotechnology precipitates a period of disequilibrium. Technological change leads to the need for institutional adaptation and/or the establishment of new institutions.

While technological progress is generally thought to be welfare enhancing, the process of moving to a new equilibrium will create losers as well as winners. Losers can be expected to use any means available to protect their vested interests, including the harnessing of existing institutions in support of their cause. If those institutions are faced with a loss of influence (and resources) as a result of the changing circumstances, there may be a natural alliance between vested interests facing diminution of economic opportunities and the bureaucracies of increasingly irrelevant institutions. These alliances can be very effective at slowing the pace of technological change even if, historically, they have not been particularly effective at stopping it. It has always proved very difficult to stuff the genie of new technology back in its bottle. The very act of attempting to slow the changes brought by technological advances, however, can create increased tensions in the economy and society as potential winners are frustrated by their inability to act on obvious commercial opportunities or consumption alternatives.

This is not to suggest that the potential winners from the advent of a new technology cannot use existing institutional arrangements to their advantage. If the regulatory regime is biased toward the acceptance of new technology or, more likely, has failed to anticipate facets of the new technology's effects, then existing regulatory processes may not be sufficiently robust to provide an appropriate balance between facilitating the introduction of the technology and the exercise of a judicious degree of caution. In these circumstances, those whose interests are served by existing international institutions can be expected to resist their reform. As a result, there may be an interest by those less well served by those same institutions in creating new institutions that better serve their needs. The result is conflicting international institutions and regulatory confusion that raises costs for those wishing to engage in international transactions involving products arising from the transformative technology and, more importantly, increases the level of risk so as to inhibit investment in further advances in, or applications of, the technology (Kerr, 2002).

When governments perceive that their country's future relative prosperity depends to a considerable degree on the successful exploitation of transformative technologies, constraints that inhibit investments in those technologies become highly politicised issues that can set the stage for major international confrontations (Isaac, 2002). Biotechnology, through its commercial use of genetic information, is a central component of the 'knowledge economy', along with the widespread use of computers for information processing and the transaction cost reducing opportunities for sharing information created by the Internet. Many developed countries have embraced the fostering of the knowledge economy as the foundation of their industrial strategy. Their focus on the knowledge economy is expected to extend well into the 21st century (Boyd *et al.*, 2003). Hence, disputes over international regulatory reform or jurisdictions are likely to be played out at the highest political level. They will not be left to the quiet backrooms of professional diplomacy. All of this means that international institutions are likely to be particularly ineffective during the disequilibrium produced by a transformative technology. Consensus-building will be difficult, if not impossible. Compliance with previously agreed international norms is likely to fall off putting considerable stress on the relations among the sovereign states that are members of international institutions.

Filling the International Regulatory Void

A transformative technology represents a fundamental change. This means that there will be major gaps in the regulatory architecture both domestically and internationally. These gaps may range from the absence of a common vocabulary that can be used to describe the technology's many facets, to customs classifications for the new goods created, to methods for assessing its risks. Hence, in the period that follows the first commercial application of a technology there will be considerable international activity that attempts to complete the regulatory architecture. Only as the architecture nears completion will the major areas of disagreement become apparent. In the early period there may be considerable duplication of effort, jurisdictional confusion and conflicting outcomes as international organisations independently take the initiative to fill the void. There may also be private sector initiatives in international standard setting and trade facilitation (Phillips and Kerr, 2002).

At least seven international bodies have been vying to co-ordinate and regulate the food and environmental safety of biotechnology products (see Table 6.1). Conceptually, they represent a progression from institutions that are largely science-based—IPPC, OIE, Codex—to ones that have broader objectives such as trade facilitation, environmental protection and other social and political goals—OECD, regional initiatives, the WTO and the Cartagena Protocol on BioSafety (BSP).

The IPPC and OIE are multilateral treaties that seek to protect plants and animals from the spread of pathogens through international trade. The IPPC protects natural flora, cultivated plants and plant products and the OIE protects animals and fish. In collaboration with both regional and national plant and animal protection organisations, they provide a forum for international co-operation, harmonisation and technical exchange of plant and animal protection information. The IPPC has addressed the international regulation of GM crops through several ISPMs while the OIE has developed standards for diagnostic reagents, sera, and vaccines for animals in the International Animal Health Code. Both institutions have their own non-binding dispute avoidance and settlement systems, but their most important role in international trade policy is through the WTO's Sanitary and Phyto-sanitary Agreement (SPS), which accepts the IPPC and OIE standards as the base for evaluating SPS disputes. National measures based on international standards from either of these institutions will generally not be open to challenge under the WTO dispute resolution process. Furthermore, both organisations nominate experts for SPS dispute panels at the WTO and provide technical background information to the panels. As such, they can have far-reaching economic and political consequences for international commerce in the products of biotechnology.

The Codex Alimentarius Commission provides a similar service related to processed foods. The Codex develops international food standards that identify the product and its essential composition and quality factors, identify additives and potential contaminants, set hygiene requirements, provide labelling requirements and establish the scientific procedures used to sample and analyse the product. Each standard normally takes six or more years to develop.

Table 6.1: International institutions regulating international trade in GM crops

Institution	Date	Coverage	Member states (2003)	DSM	Orientation
International Office of Epizootics (OIE)	1924	Infectious animal diseases	164	Non-binding; sets WTO standards via SPS S.3.4	Harmonises import and export regulations for animals and animal products through International Animal Health Code
WTO/GATT	1947	Trade in all goods and most services	146	Binding	Establish rules for transparency and dispute settlement through TBT and SPS agreements
International Plant Protection Convention (IPPC)	1952	Pests and pathogens of plants and plant products	125	Non-binding; sets WTO standards via SPS S.3.4	International standards for plant measures involving quarantines
OECD	1961	Harmonisation of international regulatory requirements, standards and policies	31	None	Consensus documents
The Codex Alimentarius Commission (Codex)	1962	Food labelling and safety standards	168	Non-binding; sets WTO standards via SPS S.3.4	International standards to provide guidance for the food industry and protection for consumer health

Table 6.1: International institutions regulating international trade in GM crops
continued

Institution	Date	Coverage	Member states (2003)	DSM	Orientation
Regional initiatives	1990s	Harmonisation of the science of regulation	Bilateral	None	Regional side agreements, MOU, MRA, formal dialogues, and joint research projects
BioSafety Protocol	2003	Transboundary movements of genetically modified organisms	73	None	Required 50 countries to ratify before it became operational – condition satisfied in 2003

Note: dispute settlement mechanism (DSM); memorandum of understanding
(MOU); members of regional associations (MRA)
Source: Buckingham and Phillips (2001); updated by the authors

Determination of the safety of the food product is based on scientific risk analysis and toxicological studies. Once a Codex standard is adopted, member countries are encouraged to incorporate it into any relevant domestic rules and legislation but they may unilaterally impose more stringent food safety regulations for consumer protection, provided the different standards are scientifically justifiable. Codex plays an important role in agri-food trade because its standards, guidelines and recommendations are acknowledged in the SPS and Technical Barriers to Trade (TBT) agreements of WTO. As with the IPPC and OIE standards, domestic standards based on Codex standards are expected to be sustained if trade disputes arise. While there are currently no Codex standards in place for products of biotechnology, there has been significant effort in Codex to develop a standard for labelling food products derived from biotechnology.

The OECD has actively assisted in the harmonisation of international regulatory requirements, standards and policies related to the discipline of biotechnology since 1995. The OECD has undertaken a number of projects to make regulatory processes more transparent and efficient, to facilitate trade in the products derived through biotechnology, and to provide information exchange and dialogue with non-OECD countries. The OECD leads efforts to develop Consensus Documents, a set of scientific background documents mutually recognised by member states that set out the biology of the crop plant, introduced trait, or gene product, and to provide a common base to be used in the regulatory assessment of an agricultural or food product derived through modern biotechnology. In addition, in June 1999 the G8 group of major economies requested that the OECD expand its work on the implications of biotechnology for food and environmental safety.

A number of bilateral or multilateral regional initiatives have played, and will increasingly play, an important role in the regulation of trade in goods and services, as noted in Chapter 5. The Trans-Atlantic Economic Partnership between the US and EU, for example, has undertaken talks in recent years to improve regulatory processes and scientific co-operation. The EU has similar trade liberalisation initiatives with Canada and Japan and since 1998 the Canadian Food Inspection Agency and the United States Department of Agriculture have also been working to harmonise their regulatory review processes. Regional agreements, Memoranda of Understanding, mutual recognition agreements, formal dialogues, and joint research projects are mechanisms that can be used to decrease bilateral regulatory barriers to GM food trade. Using these mechanisms may help countries achieve the greatest degree of trade liberalisation that is possible at a given point in time.

The World Trade Organisation (WTO) has become a major focal point for resolving biotechnology disputes. While the WTO agreement permits national standards or regulations for the classification, grading or marketing of commodities in international trade (Article XI), and the adoption or enforcement of measures necessary to protect human, animal or plant life or health (Article XX(b)), it sets some limits. The SPS Agreement specifies that: (1) measures should not discriminate between countries; (2) standards which conform to international standards developed by international organisations (i.e. Codex, OIE, and IPPC) are presumed to be consistent with the obligations outlined in the SPS Agreement; (3) standards that are in excess of established international standards or where no international agreement exists must be based on scientific principles and the completion of a risk assessment; and (4) measures shall not constitute a disguised restriction on international trade.

If a country believes another party has unjustly restricted market access through an SPS measure, it can use the WTO dispute settlement system. To date, WTO panels have decided three cases concerning the validity of national SPS measures: the EU hormones case; the Australian salmon case; and the Japanese agricultural products case. In all three cases, the contested domestic SPS measure was struck down on the basis that there was no risk assessment completed to support the SPS measure or the risk assessment was improperly done. The salmon case decision sets out that a proper risk assessment: (1) must identify the disease(s) a Member wants to prevent within its territory and the potential biological and economic consequences associated with the entry or spread of the disease(s); (2) must evaluate the likelihood of disease entering and causing damage without the measure; and (3) must evaluate the likelihood of the entry or spread of the disease with the SPS measure (WTO, 1998, par. 121). If the risk assessment does not even refer to the SPS measure, it would appear to be doomed to failure before a WTO Panel.

While the WTO is an attractive site for dispute resolution for many countries, it has some problems. As currently interpreted, the SPS Agreement does not permit non-science concerns such as consumer preference, animal welfare, or non-measurable environmental risks to be considered in the determination of whether a SPS measure is acceptable (Gaisford *et al.*, 2001). WTO dispute cases show that the organisation is likely to value unrestricted trade and scientific proof above other factors such as environmental protection or socio-economic considerations; there are no provisions to allow protection based on consumer preference. There is a very real risk that WTO decisions that are contrary to domestic concerns may be ignored, which could topple the WTO Panel process and threaten the WTO as a whole, jeopardising the other operations of the WTO in trade liberalisation. In the beef hormone case between the US and Canada on one side and the EU on the other, the EU (which lost the case) has chosen to ignore the WTO panel and accept retaliation. Accepting retaliation signals a breakdown in the political compromise and the need for renegotiation (Kerr and Hobbs, 2002). The EU has signalled that it wishes to renegotiate the SPS to take account of consumer preferences (Kerr, 1999). If an international regulatory regime is going to evolve to accommodate life science products, the conflict between a science-based regime and consumer preferences will have to be resolved (Perdikis and Kerr, 1999).

Finally, the BioSafety Protocol (BSP) represents a recent effort to provide a comprehensive international structure to ensure the protection of biodiversity and to facilitate consideration of non-scientific concerns. The BSP is a new international institution negotiated specifically to deal with trade in the products of biotechnology. The Cartagena Protocol, the formal designation of the BSP, concluded in Montreal in January 2000 provides rules for transboundary movements of GM organisms intended for environmental release and for those destined for the food chain. For living GM organisms (e.g. seeds for propagation, seedlings, fish for release), exporters will be required to obtain approval from importing countries. Within 15 days of approving a new GM variety, a country would notify a BioSafety Clearing House with information about the traits and evaluations. The first time that new GM variety is to be exported as seed, the exporting country would notify the importing country. The importing country would then decide whether to approve the shipment or decline the shipment because of risks identified through a science-based risk assessment. This process is called 'advanced informed agreement' (AIA). Although this seems straightforward, the protocol includes two features that may raise conflict in coming years. First, the text indicates that countries may in their reviews of GM foods consider socio-economic factors (e.g. the impact on local farmers), provided they respect their other international obligations. Second, the protocol specifically includes the 'precautionary principle', whereby countries do not have to have complete scientific certainty to block imports of a GM product that they fear could be harmful to biological diversity. As with the SPS Agreement under the WTO, temporary bans may be permitted but it is likely that countries will need to make real efforts to undertake the scientific research to validate (or refute) the concern. Framers of the BSP have attempted to focus it tightly on environmental risks. To that end, transboundary movements of genetically modified organisms intended for food, feed and processing (e.g. commodities) will be exempt from the AIA provisions. Nevertheless, exporters must label shipments with GM varieties as 'may contain' GM elements and countries can decide whether to import those commodities based on a scientific risk assessment. GM varieties intended for contained use (e.g. national breeding programmes and research) and those in transit through other countries will not require AIAs.

While individuals have accepted or rejected genetically modified products, or are trying to sit on the fence, societies have also collectively arrived at different positions regarding the technology that makes biotechnology products possible. These different collective conclusions have led to domestic regulatory systems that diverge considerably. Given the slowness with which international institutions are able to make progress in accommodating the types of changes brought about by a transformative technology, national governments will have to move more quickly to put in place domestic regulatory regimes – to assure those who feel threatened by the changes brought by the technology that there is order in the world rather than the chaos that they fear. Once having arrived at divergent domestic regulatory outcomes to deal with the transformative technology, national governments will seek out international organisations that most closely mirror their domestic regimes and then try and mould them so that they more closely reflect those regimes. Part of this is timing. Governments will become sensitised to issues at different points of time and there may be advantages in being a 'first mover' in the international regulatory arena if one can get one's particular vision on the table before other countries have realised the transformative nature of the technology or that it is going to be a divisive policy issue.

First mover advantage may allow a country, or group of countries, to capture an international organisation. Once an international regulatory architecture is configured in a particular way, it will be difficult for other countries to have it reversed because of the consensus nature of decision making. If the transformative technology is a sufficiently divisive issue, faced with an international regime that does not reflect their domestic policy conclusion, governments may seek to establish an alternative international institution whose architecture can be moulded to more closely reflect their own. This is clearly the case with biotechnology where the US finds the WTO closer to its domestic model for regulating biotechnology. It has been able to successfully block all EU attempts to have the WTO reformed by refusing to agree to have EU issues put onto the agenda for negotiations. The EU has tried to have the SPS opened for renegotiation on the precautionary principle and to have biotechnology dealt with in the negotiations on agriculture, both to no avail. Clearly frustrated with their inability to have the WTO modified and dismayed (but probably not surprised) with the results of the Panel decision in the beef hormone case, which is often viewed as a test case for the products of biotechnology (Kerr and Hobbs, 2002), the EU has set its sights on the BSP as an alternative international regulatory architecture. International law is sufficiently imprecise on the issue of conflicting international treaty obligations for it not to be clear whether a precedent among organisations can be established (Isaac, 2002).

Governments have an interest in having international organisations closely reflect their national regulatory architecture on important issues such as transformative technological change. Having to conform to an international norm when the existing domestic regulatory regime does not align well with it can pose considerable political risks. If the transformative technology was controversial and the national legislation put in place (at least, in part, to allay the fears of civil society groups worried about its chaotic effects) forces a government to 'lower' domestic standards to conform to externally imposed international norms, then a government could be open to claims that it has surrendered too much sovereignty to 'faceless bureaucrats' in far off places and is putting its citizens at risk. This can be a particularly emotive issue when food safety or the environment is involved—as is the case with biotechnology. On the other hand, if the international regime is strict relative to the domestic commercial environment, conforming to the international norms will reduce profitable opportunities. Firms that made investment decisions based on the set of domestically imposed constraints will find those investments threatened and others will see new profitable opportunities disappear. Politicians will be accused of pandering to foreign interests to the detriment of domestic economic growth and jobs. Faced with either prospect, governments rather than choosing to alter their domestic regulatory regimes may attempt to create an alternative international institution that more closely conforms to its domestic regime. Of course, this only delays the issue until the inevitable clash of competing international institutions.

In the case of biotechnology, two radically different international institutions exist that are to govern international commercial relations in the products of biotechnology. These institutions correspond roughly to the transatlantic regulatory regionalism that currently exists for biotechnology (Isaac, 2002). The US and other countries that have established regulatory regimes that have allowed the commercialisation of biotechnology to flourish find the existing WTO architecture to their liking and have successfully blocked any attempts to have it altered to more closely reflect the domestic regulatory regime of the more cautious EU and other countries that are less enthusiastic regarding the technology. On the other hand, the EU has worked hard to get the BSP ratified. It required the ratification of 50 countries before it came into force. This was accomplished in the summer of 2003 and it came into force in September 2003. Thus, despite the plethora of international organisations that attempted to fill the regulatory void created by biotechnology, two have become the focal points of regulation for the transformative technology. Moves have already been made by the US and Canada to have the WTO decide which regulatory regime will take precedence (Isaac and Kerr, 2003). Thus, it is important to understand the divergent approaches of the two organisations and, in particular, how they deal with the issue of liability.

Liability and International Institutions

In a world characterised by perfect, costless information and where adjudication institutions could operate without imposing transaction costs on those who chose to enter into a transaction, the existence of the threat of liability would lead to an efficient market outcome. In other words, neither party to a transaction could be surprised by the outcome of the transaction and all parties external to the transaction could not suffer a loss due to a negative externality arising from the transaction. Perfect information would mean that both parties entering into the transaction would have full knowledge of the outcomes arising from the transaction including outcomes that would arise in the future. All parties external to the transaction would know about the transaction and also understand any costs imposed on them as a result of the transaction.

The assumption of costless information means that no one, whether or not they were entering into the transaction, would have to expend resources on acquiring information pertaining to the effects of the transaction. The existence of legal liability would then act as a perfect deterrent as the full costs of any transaction would have to be accounted for by those entering into the transaction. Liability would ensure that only profitable transactions were entered into after fully accounting for compensating anyone who might be negatively affected by the transaction. Liability would, at times, serve a deterrent function by raising the cost of entering into a transaction to the point where it no longer appears profitable.

Of course, for liability to fulfil its deterrence/compensation role perfectly requires that the institutions that adjudicate and enforce liability claims operate without cost. This means that the adjudication institutions have costless access to the same information as the parties to the transaction and those external to the transaction. As a result, those who could be negatively impacted by the transaction would have the credible threat of litigation and have the assurance that they would be fully compensated if the transaction was to go ahead.

The theoretical characterisation of the role of liability developed above corresponds to the role liability plays in a transaction under the normal assumptions of neoclassical economics (Hobbs and Kerr, 1999). If the assumptions of the neoclassical model hold there would be no need for government regulatory intervention in the economy. If there is perfect, costless information about events now and in the future there can be no market failures that require government intervention to correct. Perfect costless information combined with a legal system that imposes no transaction costs on the parties that might wish to use it means that all parties would be fully compensated and, hence, there would be no role for government to intervene to reduce situations where individuals would not receive full compensation from the market or to mitigate the losses of those who fail to receive full compensation through private liability.

The neoclassical transaction is only a theoretical benchmark, similar to a perfect vacuum in physics, whose assumptions can be relaxed to more closely reflect the real world. While the assumptions of the neoclassical transaction can be relaxed to gain insights into real world effects, models can never be constructed so as to fully represent the real world, so that policy makers can never be fully informed regarding the markets where intervention may be justified or the complete ramifications of their actions. Further, as with market participants, governments do not have access to perfect costless information so that their interventions can only be approximations, at best. Government intervention is also not costlessly implemented meaning that policies cannot be precisely targeted at each transaction but rather tend to be broad brush strokes. The result is that market failures cannot be fully offset. This means that governments may choose not to intervene in some market failures as their broad brush strokes may lead to an outcome that deviates from the optimum to a greater degree than allowing the market failure to take place. In other words, governments may tolerate some market failures. If transactions are international, governments may have given up some of their powers to intervene to the international organisation.

Biotechnology, as well as other transformative technologies, has inherent characteristics that mean that transactions take place under conditions that deviate considerably from the neoclassical assumptions. The newness of the technology means that rather than being perfect or costless, information on possible externalities is incomplete and costly. This is particularly true in the case of the distant future regarding the human health effects of consumption or potential damage to the environment. Given long lead times and imprecise measurement, it may not be possible to identify with any degree of surety the source of a negative externality. This means that liability cannot be accurately assigned and, as a result, it cannot fulfil its deterrence/compensation role. Of course, the adjudication system imposes considerable transaction costs on those that wish to invoke its use. Further, it does not operate in a predictable fashion. This is particularly true in the case of international transactions where no international private commercial law exists.

In situations where liability cannot effectively serve its deterrence/compensation role, governments may choose to intervene to correct market failures.[1] Intervention in cases of transformative technologies is more often of a regulatory nature than handled through taxation. If liability cannot efficiently serve its deterrence/compensation role, then governments may intervene to replace the disincentive of deterrence with regulatory hurdles or prohibitions. If firms find the regulatory hurdles too costly, investment in developing the technology will be inhibited. If properly designed, regulatory hurdles and/or prohibitions can protect individuals from uncompensated losses that may arise from transformative technologies.

There is one group that may lose from the changes brought by transformative technologies that society has most often chosen not to

---

[1] Governments may also choose to intervene when private liability acts as a deterrent to the development of a transformative technology that it wishes to promote. In this case the government may feel that incomplete information would set premiums for insurance against liability too high relative to what they would be once experience was gained in the development of the technology. This is certainly the case in the nuclear power industry where governments have passed legislation that allows nuclear generators to be underinsured, effectively limiting their liability. This was the only way investment could be attracted to the industry (Kwaczek *et al.*, 1990).

compensate. These are competitors with the products arising from the transformative technology that continue to use the less efficient previous technology. Replacing the inefficient with the efficient through competition is seen as economic progress and welfare enhancing for society. The economic model that underlies this view of technological change is based on the premise that consumers always win from the lower prices arising from increased efficiency and that their gain outweighs the losses of inefficient producers. As the gain to society is greater than the loss it is theoretically possible for the winners to compensate those who lose and still be better off (the compensation principle). Whether or not compensation takes place is not considered as being important, only that compensation could take place. This does not mean that inefficient producers will not seek protection from governments to prevent their loss. Of course, this is the basis of the protectionism that is the heart of trade policy and which organisations such as the WTO have been put in place to manage.

The debate over globalisation is often couched in terms of certain groups; e.g. small traditional farms, losing out to the large modern farm enterprises that adopt the transformative technology. As they have always done since the time of the Luddites, losing groups (or their apologists and advocates) attempt to have themselves endowed with attributes that suggest they are relatively more valuable to society than the winners. This is done to convince society that it will actually suffer a loss as a result of their disappearance as significant economic entities. Sometimes they are successful.

It is not common, however, for winners to be liable for the economic losses suffered by other groups in society as a result of transformative technological change. Car manufacture's were not considered liable for the losses of those who had previously been engaged in the manufacture of horse-drawn buggies. The makers of steamships were not liable for the losses of sail makers. Firms offering cell phone services are not liable for the losses of those who provide land-line telephone services. Governments may, at times, be willing to step in and offer groups protection from losses that arise from the forces of competitive technological change, but these interventions need to be put in a separate category from those where governments intervene when liability would apply if there were perfect costless information and the adjudication system operated without transaction costs.

Faced with a market failure in the liability system, governments have attempted to intervene in the markets for biotechnology through regulation. The pace at which the transformation is taking place as a result of biotechnology has outstripped the pace at which international organisation can adapt to accommodate the new technology. As a result, national governments have had to act independently in establishing their regulatory regimes. Predictably, they have arrived at different conclusions regarding the nature of the market failure in the liability system and the degree of failure. In particular, the US and the EU have radically different views of the extent of the market failure. These differing domestic views of the market failure shape their response to the international regulation of biotechnology.

The major difference in their perception of the market failure in the liability system relates to the effects on human health from long-term consumption of the products of biotechnology and the long-run effects of releasing transgenic organisms into the environment (Gaisford *et al.*, 2001). There are no important differences in their approach to the short-term human health effects from consumption or short-run impacts on the environment. In the EU, however, there is some confusion between market failure in liability and protection of those who would lose from the introduction of the technology. The inclusion of protectionism in the EU's regulatory approach to biotechnology particularly irks the US because it is simply seen as a further extension of the agricultural protectionism of the EU's Common Agricultural Policy (CAP). The EU and the US have been locked in a long-term battle over EU agricultural policies and how they distort international markets (Gaisford and Kerr, 2001). Producer protectionism, however, is a side issue in the transatlantic regulatory debate over how to deal with the concerns of consumers and other members of civil society who are sceptical regarding long-term human health and environmental effects of biotechnology (Gaisford *et al.*, 2001).

When it is perceived that liability is prevented from effectively fulfilling its deterrence/compensation role, the most common regulatory alternative is to act to reduce the risks that may arise from consuming the product or from its release into the environment. Typically, governments require that firms wishing to sell their products in the marketplace satisfy certain regulatory standards before the product can be sold. Often, this involves testing and providing other scientific evidence for review by experts appointed by the government. This is an attempt to replace the deterrent role of liability with a science-based system for product approval. If the product clears the regulatory hurdles, then there is no need for compensation through liability because a problem will not arise. Passing the regulatory hurdle usually means that firms are absolved from liability if all of the information provided was accurate. Firms that are caught cheating on regulatory procedures are subject both to prosecution by the government and to private liability. Problems arise, however, when firms follow all the regulatory procedures and subsequently there is found to be a problem with the product. The use of asbestos in building materials and the drug thalidomide are two obvious examples. Thus, government regulatory actions can only replace the role of liability to a partial extent.

If the technology is new there may not be sufficient scientific information upon which to determine if the absence of liability can be offset by a set of regulatory hurdles. When incomplete scientific information exists, governments may ban products that have used the new technology in their production until sufficient scientific understanding has been acquired. Faced with incomplete information, governments may wish to act with precaution. Of course, many products fail to pass the regulatory barriers and are never allowed to be released commercially.

The heart of the regulatory divergence in domestic policy across the Atlantic lies in the operationalisation of precaution—how lack of scientific certainty is dealt with. In order to understand the differences in regulatory approaches it is necessary to understand the broad regulatory framework that governs technology. The Risk Analysis Framework (RAF) was developed to deal with the regulation of advanced technology products (which were characterised by a large information gap between the producers of the innovation and the intended consumers) where the goal was to credibly inject science into public policy development (National Academy of Sciences, 1983). The language of risk analysis is found in regulatory guidelines for the research, development and commercialisation of advanced technology products (including biotechnology) in many countries (including both the EU and the US) and in various multilateral agreements treaties and organisations (including those of the United Nations, e.g. the World Health Organisation and the Food and Agriculture Organisation).

Significantly different views on how to actually operationalise the RAF in North America relative to the European Union arose in the case of biotechnology. As a result, two quite distinct regulatory trajectories exist for biotechnology—the scientific rationality and the social rationality trajectory (Isaac, 2002). The differences between the scientific rationality perspective and the social rationality perspective begin with a fundamental difference in belief about the appropriate role of science and technology in society. According to the former, technology yields innovations and enhances efficiency thereby producing economic growth. Rising incomes lead to demands for stringent social regulations in areas such as food safety and environmental protection. The result is a regulatory race to the top fostered by technological progress (Isaac and Kerr, 2003). The goal of the regulatory regime for science and technology is maximisation of technological gain, subject to achieving acceptable standards of safety.

The social rationality perspective views science and technology as normative activities that by their very nature bring change to what is a delicate social balance of the preferences and concerns of all constituents. As change disrupts the balance, the social rationality perspective supports regulatory policies that ensure technological precaution, so that all of the impacts arising from a change are dealt with in a socially responsive manner (Isaac and Kerr, 2003). Progress in science and technology cannot be left to market forces.

Differing treatment of precaution is central to the regulatory approaches based on scientific and social rationality. While both interpret the precautionary principle as meaning that in the face of scientific uncertainty, regulators must employ precaution, the two perspectives operationalise precaution in fundamentally different ways.

The scientific rationality perspective operationalises the precautionary principle as a risk assessment tool. It is the scientists entrusted with risk assessment that decide when the dearth of information justifies a go-slow (or stop) with a new technology. When evaluating a new technology there is, of course, an absence of data on the risks and, hence, risks are calculated according to causal-consequence models built from the accumulated peer-reviewed scientific literature (Isaac and Kerr, 2003). Hence, as a scientifically rational risk assessment tool, the exercise of precaution is grounded in sound science where risk assessors who hold a required amount of scientific credibility are the ones who are charged with making decisions. A rules-based approach is the result.

The social rationality perspective goes further and operationalises the use of precaution as both a risk assessment tool and also as a risk management tool. According to this perspective, risk assessors can use the precautionary principle in the same way as in the scientific rationality approach but it can also be legitimately employed by risk managers to ensure precaution in the face of non-scientific perceptions and concerns. Using the precautionary principle as a risk management tool potentially increases the social responsiveness of regulations but also increases the discretionary nature of regulations. As it has a broader social mandate it includes socio-economic considerations. This leads to the possibility of regulatory steps to prevent the technology imposing negative economic consequences on those who cannot or will not use the new technology and their broader communities. Implicitly, then, it opens the door to compensation for those who will lose as a result of the technological change. When international trade is involved, it leads to trade barriers that are not fundamentally different from traditional protectionism.

The World Trade Organisation and the Biosafety Protocol

Out of the initial plethora of international organisations with an interest in biotechnology regulation, two have emerged as the primary regulatory bodies. As currently structured, these loosely conform to the scientific rationality approach and the social rationality approach and there is now fierce competition over which will become the ruling paradigm for international trade in the products of biotechnology. The World Trade Organisation embraces the scientific rationality approach while the Biosafety Protocol embodies the social rationality approach. The BSP has been able to develop in the form it has because, due to a quirk of history, the US is not able to become a signatory and, thus, cannot formally participate in its framing. The BSP comes under the auspices of the United Nations Environment Programme's Convention on Biological Diversity (CBD) which arose out of the UN's Earth Summit in Rio de Janeiro in 1992. The US did not sign the CBD and, hence, cannot formally be part of initiatives undertaken under its auspices. As a result, the European Union was able to influence the protocol's design without official US opposition.

It should be noted that the WTO's scientific rationality approach predates the point in time when biotechnology became an important international issue. In the Uruguay Round of General Agreement on Tariffs and Trade (GATT) negotiations, there was a major initiative to bring trade in agricultural products into conformity with general GATT principles. Prior to this, most agricultural trade was conducted under 'waivers' that allowed the use of non-tariff measures to restrict imports and the unrestricted use of subsidies. With stronger disciplines on the tools of agricultural protectionism likely to arise from the Uruguay Round negotiations, there was a worry that demands for protection for agricultural products would be increasingly satisfied through the nefarious use of border regulations justified on human, animal and plant health grounds. To forestall this circumvention of the increased disciplines on trade in agricultural products, a new Agreement on the Application of Sanitary and Phyto-sanitary Measures was negotiated. It was given a scientific rationality basis to remove it from the political process, thus hopefully isolating it from protectionist influences. The US and the EU accepted this (Kerr, 2003b).

The WTO is a trade agreement. Its primary purpose is to limit the ability of governments to impose trade barriers—so that those who invest in international trade activities do not have those investments threatened by the capricious use of trade barriers. It also has a long-term goal of trade liberalisation. The economic model that underlies the WTO is neoclassical in nature. In particular, it assumes that consumers have perfect costless information and, hence, they always benefit from trade liberalisation because prices fall as a result of the removal of trade barriers (Isaac *et al.*, 2002). They also benefit from open markets because the efficiency benefits of technological change are translated internationally into lower prices. Consumers have no reason to ask for protection and thus, their asking for protection was not anticipated by the framers of the WTO. As low international prices have a negative impact on producers, they can be expected to seek protection. Thus, the entire structure of the WTO is aimed at controlling producer-based attempts to receive protection from their governments.

The result is that the WTO does not recognise the market failure that can arise from incomplete information regarding a new technology (Gaisford *et al.*, 2001). It has no mechanism to accommodate consumer or environmentalist requests for protection from new technologies such as biotechnology. Beyond the failure to recognise that such a market failure exists, there are considerable practical problems that must be solved before the WTO can incorporate consumer and other groups' demands for protectionism, including how such claims would be legitimised and how to prevent them from being captured by producer protectionist interests (Perdikis and Kerr, 1999). From the point of view of an international trade agreement, the potential harm is probably greater than the potential gains. Thus, the WTO exhibits considerable inertia on this issue. This inertia is now championed by the US to prevent consumer issues from being used to inhibit its exports of products of biotechnology. It has refused to allow the issue to be put on the negotiating agenda despite EU requests to open up the SPS for re-negotiation so that these issues could be dealt with. While the SPS may not be the best place to deal with these types of market failure (Perdikis and Kerr, 1999), they have not even been allowed to become an agenda item in the SPS or in any other WTO negotiating forum.

Under WTO commitments, a country cannot impose trade barriers based on long term human health or environmental reasons unless the product fails the scientific risk assessment test. The US and Canada have challenged the EU's temporary moratorium on imports of GM products at the WTO and look likely to challenge the new EU labelling regime. Given the current WTO rules, they are likely to win (Isaac and Kerr, 2003). This will be a re-affirmation of the WTO system based on scientific rationality. This means that the WTO acts as if no market failure exists that governments can legitimately intervene to correct. Thus, implicitly, the WTO is acting as if the deterrence/compensation role of liability is able to fulfil its role. In other words, it does not have to directly address these issues because the threat of liability is sufficient to prevent the commercial release of an unsafe biotechnology product or, if one is released, those negatively impacted will be adequately compensated. This deterrence or compensation will have to be accomplished through private commercial law. It is well known, however, that private international law suffers from large transaction costs, difficulties with conflicts of law and far from transparent outcomes (Wasylyniuk *et al.*, 2003). Further, given the difficulties in pinpointing the source of a long-term human health problem or environmental degradation, relying on liability to provide the correct degree of deterrence or compensation seems problematic. It is, however, consistent with the neoclassical tradition of the WTO. To be fair, the WTO is a trade agreement and does not have a mandate that extends to private international law. Any improvement to the handling of liability in international law would have to be undertaken in a different venue.

The BSP, in contrast, starts from a different premise and is firmly rooted in social rationality. The BSP accepts that a market failure can exist and that countries have a precautionary duty to attempt to correct that market failure. According to Isaac *et al.* (2002):

> The Biosafety Protocol has as its primary focus the protection of the natural environment—although ... the definition of 'natural environment' is very broad and includes the economic well-being of certain groups. In the Biosafety Protocol, governments have a responsibility to behave proactively to prevent potential market failures that might arise from transboundary shipments of LMOs [living modified organisms]. The protocol allows governments to meet this responsibility through the use of trade barriers, if necessary. (p. 106)

Central to the social rationality approach to precaution is that the ultimate responsibility for its invocation is political rather than being vested in a scientific body, as is the case with science rationality-based regulatory systems (Kerr, 2003b). Science can inform the decision in the case where social rationality is applied but the ultimate decision is a political one (Isaac and Kerr, 2003). From a trade policy perspective, there is no guarantee that political decision makers will deal only with questions of scientific uncertainty—they can also be swayed by traditional protectionist interests.

The European Union has tried to deal with how politicians should operationalise the precautionary principle. Some environmental NGOs and other civil society groups opposed to biotechnology hope that what van den Belt (2003) calls the strong version of the precautionary principle will be the operational definition. The strong version, they believe, can be used Luddite-like to deny the technology. Van den Belt (2003), however, shows that the strong version of the precautionary principle is logically inconsistent. The EU has not accepted the strong version of the precautionary principle but has been unable to articulate the operational principles under which it can be applied and, hence, decision-making lacks transparency (European Union Commission, 2000). Hence, the precautionary principle was enshrined in the BSP before its major proponent had been able to determine how it is to be operationalised.

This means that countries can invoke the precautionary principle under the BSP and exclude products of biotechnology from their market for political reasons. Consistent with the social rationality approach, in Article 26 of the BSP specific provisions are made for socio-economic considerations to be taken into account when assessing a product's suitability for importation (Phillips and Kerr, 2000). A further flaw in the BSP is that there is no binding dispute settlement procedure (Kerr and Hall, 2004). This means that countries that abide by the BSP rules can, if they believe that liability is not able to efficiently or effectively play its deterrence/compensation role, act with impunity to correct the market failure using trade measures.

The framers of the BSP, however, understood that existing international liability mechanisms were unlikely to be able to efficiently or effectively perform their deterrence/compensation function and issues of liability were included on the negotiating agenda. The negotiating parties could not agree on how to deal with liability, however, and it was agreed that the issue would be dealt with through further negotiations. Article 27 of the Protocol states that:

> The Conference of the Parties serving as the meeting of the Parties to this Protocol shall, at the first meeting, adopt a process with respect to the appropriate elaboration of international rules and procedures in the field of liability and redress for damage resulting from transboundary movements of living modified organisms, analysing and taking due account of ongoing process in international law on these matters, and shall endeavour to complete this process within four years.

It seems clear that some of the negotiating parties thought the issue of liability was very important and that if the Protocol was going to be effective it needed to have a specific liability regime, and relatively quickly—hence the relatively short four-year time frame. The group charged with implementing the BSP work plan—the Intergovernmental Committee for the Cartagena Protocol on Biosafety (ICCP)—immediately began to collect information both from international liability experts and from individual Member States of the BSP regarding their domestic liability regimes for biotechnology. On reading the documents produced by the ICCP or under its aegis it seems clear that it was hoped that an existing international liability regime could be adopted, or at least easily adapted by the BSP.

A range of existing international treaties, agreements and arrangements were investigated by the ICCP in search of a model, including a number of nuclear-liability treaties, oil pollution liability instruments, arrangements governing the transport of dangerous goods and substances, the 1972 Convention on Liability for Damage Caused by Space Objects, the Convention on Civil Liability for Damage Resulting from Activities Dangerous to the Environment of 1993 (the Lugano Convention), and the Convention on the Protection of the Environment Through Criminal Law, 1998 (ICCP, 2001). All of these arrangements were found to be deficient in one way or another for the case of biotechnology. While a wide range of issues were raised, the major deficiencies related to: (1) the long time frame between the international movement of genetically modified products and when environmental damage would be detected; (2) how to value damage to biodiversity; (3) what parties would be liable; and (4) how to deal with liability when regulatory provisions had been complied with. At the time of writing, the ICCP had not resolved any of these issues satisfactorily. Given the existence of an additional wide range of less substantial issues that have been raised but not resolved, the four-year time frame looks increasingly untenable.

The long timeframe relating to the potential damage from the release of genetically modified organisms does not arise to the same degree in any of the other international arrangements relating to liability because, while damage may take place over an extended period, for example, from a nuclear accident, the problem will be immediately apparent after the nuclear accident. Most international agreements pertaining to liability have statutes of limitations built into them—the claim must be made within thirty years of the liable occurrence or within three years of when a problem becomes apparent. In the case of genetically modified organisms there is an appreciation that the effect on biodiversity may be a very long-term phenomenon related, for example, to the replacement of wild non-genetically modified species with unanticipated outcrosses of genetically modified commercial varieties. The spread of a new dominant genetically modified plant may take hundreds of years and may not even be discernible for decades. Statutes of limitations are practical compromises in liability cases. After thirty years many of the firms and other economic actors involved in the transboundary movements of genetically modified organisms may simply not exist, making it impossible to recover damages.

Liability is normally addressed by the purchase of insurance. Insurance pertaining to speculative events far in the future may be impossible to obtain because insurance firms do not know how to evaluate them. If insurance is available, it may be expensive, considerably reducing the returns of those investing in biotechnology. Firms are lumbered for years with large insurance premiums for events for which it is impossible to even establish the probability of occurrence.

If insurance is not available due to the speculative nature of the risks, an alternative is to establish a fund contributed to by the likely liable parties. Long lags between the establishment of a fund and when it may pay out means that private (and possibly public) capital is tied up for extended periods when no problem ever actually arises. This can be seen as putting an unwarranted burden on the biotechnology industry. Hence, statutes of limitation are a reasonable compromise between not unduly hampering an industry and the need for deterrence/compensation. Given the long lags expected in the case of damage to biodiversity, statutes of limitation may not be appropriate but neither may be open-ended insurance/contribution to a fund policies. The ICCP has not yet resolved this issue.

Of all the major outstanding issues, the ICCP has had the most difficulty dealing with how to value damage to biodiversity. In part, this is because legal expertise is being brought to bear on a question whose answers lie, in part, in biological science and economics. Even basic concepts relating to what would constitute damage to biodiversity have not been developed. There has been some discussion of whether a *de minimis* level of damage needs to be established to prevent large numbers of nuisance cases from being brought forward, but until the central question of what constitutes damage and how that damage can then be valued are answered, it will not be possible to establish a *de minimis* level.

As of December 2002, the ICCP had this to say about the evaluation of damage:

> There are a number of issues relating to the definition of damage. The wording of Article 27 appears sufficiently broad to encompass damage to the environment, persons or property. However, a number of provisions under the Protocol refer to [living modified organisms] LMOs that "may have adverse effects on the conservation and sustainable use of biological diversity". The Protocol does not give any guidance on how conservation and sustainable use of biological diversity would be defined.... The question may also arise as to whether "damage to the conservation and sustainable use of biological diversity" is distinct from "damage to biological diversity". Secondly, the reference "taking into account risks to human health" may bring a new dimension to the issue. In effect, personal injuries or related health costs, such as the costs of medical screening after an incident arising from LMOs, would need to be considered in any definition of damage under the Protocol. Lastly, Article 26 contemplates socio-economic considerations arising from the impact of LMOs on the conservation of and sustainable use of biological diversity. Would these considerations broaden the concept of damage under the Protocol? If so, how could this type of damage be quantified? (ICCP, 2002, p. 6)

The ICCP (2002) goes on to state:

> Another important issue is the evaluation of damage to biodiversity. In order to assess whether the damage to biodiversity occurs as a result of transboundary movements of LMOs, baseline data on the state of biodiversity must be established in advance. Moreover, evaluation criteria have to be created for the damaged natural resource to avoid disproportionate costs of restoration. If restoration is technically not or only partially possible, the valuation of the natural resource has to be based on costs of alternative solutions by establishing natural resources equivalent to the destroyed environment. (pp. 6-7)

Thus well into the ICCP's mandate there are far more questions than answers regarding the issue of defining damage and valuing it.

The questions pertaining to what is known as 'channelling liability' are also complex. According to the ICCP (2002):

> The issue of which actor should be liable for damage is one of the most critical elements in the liability regime. The majority of international legal instruments channel liability to the "operator"—the person who has the operational control of the activity at the time of an incident causing

damage.... Assigning liability to appropriate persons may be determined by principles such as fairness—reflecting equitable balance between interests of victims, environment and other stakeholders including industry; effectiveness—allocating liability to a person who is in the best position to prevent damage and purchase financial security; and transparency—facilitating the identification of the person liable. (p. 7)

In the case of transboundary movements, the operator could be defined as narrowly as the transporter or the exporter, who might be small firms and unlikely to be able to provide financial security. On the other hand, the developers of genetically modified products may not be involved at all in the transboundary movement of the product. The ICCP goes on to state, however, that:

It should also be noted that the question may arise as to whether and, if so, to what extent the liability could be channelled to actors beyond those envisioned in the Protocol, such as "developer" or "producer" of LMOs. In addition, if liability were allocated to them, clear definitions need to be developed regarding who would fall into the appropriate categories. (p. 7)

This is a clear divergence from the legal norm where the 'operator' is considered to be the liable person. It reflects the views of some participants in the biotechnology debate that those firms that are developing biotechnology products are socially irresponsible (multinational) corporations and should be held accountable for their activities. It may also reflect the perception that only they are likely to have the 'deep pockets' that could pay for restitution or compensation in the case of a major disaster. There is no precedent that the inventor, innovator or those responsible for commercialisation of a product should be held liable for problems arising from its transport or commercial use. In other words, in this view it is the very existence of a biotechnology that would create the liability rather than its use. In contrast, those who invented nuclear power are not liable in the international treaties with provisions pertaining to nuclear liability—only the operators of nuclear facilities are held accountable. Those who develop rocket delivery systems are not liable in the case of damage caused by space objects—only those who launched the spacecraft. The source of dangerous goods is not liable when there is a problem during their transport. While some member states of the BSP may want liability to be extended to those responsible for the development of biotechnology, it will be controversial and lead to difficult negotiations.

The final major area of controversy relates to liability when firms are in regulatory compliance. The BSP has regulatory hurdles that firms must pass prior to the transboundary shipment of genetically modified products being made. This includes a risk assessment by the importer. If the genetically modified organism is cleared for import under the BSP and, subsequently a problem is discovered, should the firm be liable? Further, if a firm conforms to best practices regarding the genetically modified product at one point in time but subsequently scientific understanding improves and a problem is found that damages biodiversity, should the firm be liable? The ICCP (2002) identified these two problems but provided no guidance as to how they should be resolved: "The first is whether compliance with regulatory permits would be allowed as a defence, such as approval of an LMO in accordance with AIA procedure" (p. 9).

The ICCP does note that some domestic liability systems pertaining to biotechnology do not allow compliance with regulatory permits as a defence but is silent on suggestions as to how this issue should be handled in the BSP. The ICCP (2002) goes on to ask (but not answer):

> ... whether the state-of-the art or development risk would be taken into account in applying defence, such as foreseeability or best practicable means relating to the global scientific and technical understanding on LMOs. If a risk assessment on a LMO is conducted carefully and shows that it poses no risk at the time, should liability be exempted when risks or damage emerge over time with increasing scientific and technical knowledge. (p. 9)

It seems clear that the framers of the BSP, while agreeing that existing liability systems are unlikely to be sufficient to provide a well functioning deterrence/compensation system, upon closer examination have found that it will be difficult to improve on them. Part of the difficulty relates to the nature of transformative technologies whereby scientific knowledge is going to be incomplete. Further, the nature of the technology means that very long timelines will be involved, complicating the use of liability. As a result, an effective liability regime for the BSP seems a long way off. The BSP provides a strong preventative alternative to liability given that it allows states to invoke the precautionary principle without recourse by other parties (Isaac *et al.*, 2002). However, this leaves the BSP wide open to protectionist influences. As a result, it remains a flawed instrument to implement a liability regime or its alternatives.

Conclusion

Transformative technologies represent significant events. The process of technological change is one that creates economic disequilibrium and foists change on the wider society. International institutions are no exception. International institutions are generally very slow to change. As they are voluntary, decisions must often be reached by consensus. Deliberations are often slow because there are a large number of parties involved. It is probably not surprising that no international liability regime has yet emerged for biotechnology. Two competing international organisations appear to be emerging as the primary international regulators of biotechnology—the WTO and the BSP. They are on different regulatory tracks that represent the positions of the major protagonists in the debate over biotechnology. The WTO appears unable to adapt to accommodate the international trade questions, including liability, raised by the transformative technology. In essence, it is simply acting as if no transformation is taking place.

The BSP, on the other hand, has a built-in bias against the technology both through the inclusion of the precautionary principle and the absence of a dispute settlement mechanism. While it recognises that the existing international liability system will not be adequate for liability to fully play its deterrence/compensation role in the case of biotechnology, it is finding it difficult to devise an international liability regime that deals with the characteristics of a biologically based transformative technology and that is acceptable to all of the parties to the Protocol.

The final result is that it will be some time before an international liability regime for the products of biotechnology emerges. This will lead to less than smooth international relations over the regulation of biotechnology over the foreseeable future.

Chapter Seven:

Biological Mechanisms to Control GM Liabilities

Introduction

In October 1999, Robert Shapiro, Monsanto's CEO said: "We are making a public commitment not to commercialise sterile seed technologies, such as the one dubbed 'terminator'" (Shapiro, 1999). Many viewed this statement favourably. However, costs from cross-pollination and volunteer growth of GM crops are rising. By 2005, GM wheat and new 'third generation' crop varieties will increase the need to control co-mingling of GM seeds.

Worldwide, importing nations are becoming increasingly concerned with the compositional contents of imported raw commodities. Crop purity is under greater scrutiny. When StarLink™ maize co-mingled with unmodified maize destined for the food industry, concerns about containment and purity left the backrooms and were splashed across the front pages of newspapers around the world. Governments, industry and scientists are keenly interested in finding the right mix of institutions and science to manage these risks, in order to bring forward new crops with potentially even greater value.

Present crop production practices in North America are faced with two growing concerns: cross-pollination of GM crop varieties with conventional varieties; and controlling the growth of volunteer GM seeds. Cross-pollination of GM crops with conventional or organic crops worries many producers that are trying to produce products for the organic and non-GM niche markets. Controlling the growth of volunteer GM seeds can be done chemically, but at an added cost for producers. Both of these issues need to be rapidly addressed if the adoption rate of GM crops is to remain high and the commercialisation of future varieties is to continue.

The Challenge

As suggested in the previous chapters, first generation, input-trait GM crops have for the most part been judged by regulators as substantially equivalent to existing varieties and allowed to be introduced into commodity food systems without any segregation. A number of potential liabilities, however, exist. Many of these GM crops have the potential to cross-pollinate with other crops of the same species or weedy relatives, or to become volunteers in other crops, creating potential new environmental risks that may diminish the benefits of the technologies or create quality problems and new liabilities in other crops or the food system. Second generation crops, which involve output modifications, will likely only be viable if they can assure the purity of their quality, which is problematic given the potential for gene transfer. Third generation crops, with new industrial, nutraceutical or pharmaceutical properties, will clearly require significant gene control systems, or simply will not be allowed to be produced (see Chapter 10 for a more detailed discussion). In each case, the firms are also concerned that they face dilution of some of the benefits of their innovation because of the self-reproducibility of many of the GM crops.

The first generation products of biotechnology have been with the market for about ten years in Canada, the US and several other countries. Three biotech crops account for the lion's share of GM food production: canola, maize and soybeans. For the most part, these crops entered the marketplace with minimal restricting regulations. The environment for the anticipated introduction of new generation biotech crops (e.g. involving novel uses) is vastly different from the one that first generation biotech crops faced. Later generation crops will face restructured regulatory systems, radically altered marketplaces and new technology options (Phillips and Khachatourians, 2001).

While genetic engineering has the potential to create many new GM crops that will create substantial value in the marketplace, two main challenges face the industry. On the one hand, those who are investing in these technologies are determined to capture a share of the returns in order to pay for the large development and commercialisation costs. On the other hand, firms and society are vitally interested in ensuring that the new traits and varieties created do not impose risks or liabilities that offset or swamp the value being generated. At the farm level, in particular, there is significant risk of dilution of the rents and for co-mingling of the new traits with other crops, creating potential new liabilities.

There are methods available to restrict cross-pollination and manage volunteer seeds but societal pressure would seem to have removed the option of seed sterility. Many civil society groups, environmental NGOs and developing country governments expressed concern that the 'terminator' technology could threaten landrace varieties, increase corporate concentration, reduce biological diversity and ultimately destabilise developing countries' agro economies (Visser *et al.*, 2001). In October 1999, when Shapiro, announced that Monsanto would not use 'terminator' technology, many expected that this would effectively close the door on further pursuit of this kind of gene research, as other seed development companies were expected to follow the lead of Monsanto. It has not. On March 26th, 2002, Syngenta received a patent for controlling plant fertility but since has publicly pledged not to commercialise this technology.

Fundamentally, the allocation of the benefits and management of these risks will need to be brought about by a combination of institutions and biological controls. Institutionally, the public sector will continue to have a say on when, where and how GM crops are introduced and propagated, as well as in adjudicating property rights. Meanwhile, private firms will probably have a major role in managing and enforcing contracts and systems to capture their benefits and to manage the risks and liabilities of these new crops. While regulation and markets may be able to control some gene transfers, genes from open pollinating crops are likely to continue to travel. Regardless of how effective regulations or contracts are, some actors either deliberately or inadvertently will misappropriate these new technologies, diluting the benefits and creating potential new risks and liabilities. Furthermore, even if all 'cheating' could be controlled, many plant species are promiscuous sexually, creating natural gene flows. In short, we need to look for a combination of both organisational and biological control mechanisms to manage the benefits and risks.

The Economics of GM Crops

There are two main streams of literature relevant to this issue. A considerable amount of legal and economic work exists on how patents and other intellectual property rights (IPRs) provide the conditions necessary for private investment in agri-food research. Meanwhile, much of the economics literature related to GM crops focused on evaluating the 'gains to research' has acknowledged the potential for 'externalities' that could reduce the net gain from GM innovations.

Suffice it to say, there is a large body of literature related to the role of patents in creating the incentives for private investment (e.g. Santaniello *et al.*, 2000). The establishment of enforceable property rights for products of genetic crop research has significant implications for the amount of research that the private sector will provide. In the absence of enforceable property rights, many of the products of research can be copied or reproduced. While all firms that use the research output may benefit, without property rights there is no way for the market to fully remunerate any firm for doing research. This creates a 'public good' market failure resulting in under-investment in research activities. This is one of the main explanations given for the high internal rate of return for most agri-food research, estimated by Alston *et al.* (2000) to be in the 70% range. Recently, governments, and to some extent the private sector, have addressed the 'public good' market failure in research by establishing effective property rights over the products of research. As outlined by Gray *et al.* (1999) the assignment of IPRs provide some added ability to capture value from research. Internationally the development of the International Union for the Protection of New Varieties of Plants (commonly known by the French acronym, UPOV) and plant breeders' rights has provided some support. While the increased use of utility patents for the new technologies also strengthen control, the US has gone the furthest with utility patents on new plant varieties.

The problem with the current package of IPRs is that they do not fully control the use of a new technology once it is expressed in seed. Most GM crops can be propagated in subsequent years with seed from previous years. While regulations and private contracts attempt to manage that activity, many in the industry note that they are far from effective. Even in the absence of a GM variety, industry sources estimate that in Saskatchewan alone, more than 300,000 acres of wheat in 2000 were planted to unregulated plant varieties and that up to 3% of the exports by volume may be composed of some varieties that have not been approved for release in Canada. Furthermore, officials in Monsanto have estimated that without their TUAs, they would lose as much as 25% of the royalty payments for Roundup Ready® crops without hybrids; even with the TUAs, they estimate that 10% of the acreage planted to Roundup Ready® crops may not be covered by agreements and therefore does not involve royalty payments. Clearly, risks are greater and dilution is significant when the industry has to rely simply on the power of institutions.

The second literature, mostly from economics, offers estimates of the economic cost-benefit of GM crops. While the evidence is scarce, the early estimates (e.g. Moschini *et al.*, 1999; and Kalaitzandonakes, 2003) suggest that most of the new GM crops provide fairly significant net social benefits, with the innovators capturing a large share of the returns. While most of this research has simply looked at the productivity enhancement and its impact on markets, a few researchers have examined the impact of new risks or liabilities on the demand or supply side of the analysis. On the demand side, the recent consumer backlash against GM foods highlights one possible risk of new GM crops. Following on Ackerlof's (1970) work on the market for lemons, Fulton and Giannakas (2001), for example, suggest that in some instances where consumer fears are high enough, the inability to segregate and label GM and non-GM crops and foods could result in global welfare losses. Kuntz (2001) examined the impact of GM wheat varieties on Canada's wheat exports, concluding that in a worst case, more than C$184 million or 71% of the quality premium earned in the market could be lost without effective segregation. On the supply side, a number of researchers have looked at how unmanaged risks could diminish the benefits (e.g. Mayer and Furtan, 1997). In response, a number of researchers have looked at how either government or market institutions could manage that risk. Smyth and Phillips (2001), Phillips and Smyth (forthcoming) and Lin (2002), for instance, look at how the evolution of private identity preserved production and marketing systems have evolved to manage the new GM crops. The overriding impression from the systems currently in place is that they are idiosyncratic, costly and do not manage all the concerns.

The key message from the literature is that both the distribution of benefits and the management of risks and liabilities can only partially be managed by institutions. The studies show that even with the best institutions in the world, some benefit dilution continues while risks and liabilities remain. Biological control mechanisms could provide a useful adjunct or alternative to often costly or ineffective institutional approaches.

GM Crop Liabilities

The liability cost of GM technologies escaping and going rogue or co-mingling and adversely affecting quality of other products is large and growing. A number of recent examples from the canola and other sectors showcase the impact. Three key liability issues can be identified and at least partially quantified. Both volunteers and pollen flow create the conditions that could lead to co-mingling of GM and non-GM crops, which jeopardise the value of the crop in some markets and, if undetected until it is processed into foods, entire products or product lines. Finally, inability to control gene flow has also impeded transfer of genetic material between nations that developed the new varieties and those that want to adopt new technologies.

There are two ways that GM genes flow and create liabilities. First, through normal agriculture practices, seed are left behind during harvest that germinate in the spring and, depending on the crop planted, may create a tolerance-level liability. The second is that pollen of the GM plant could fertilise a conventional plant and the resulting hybrid seed has the potential to possess the trait for that GM gene.

There is no harvesting system in place in the world that is capable of containing all the seeds that are produced on a plot of land. This is due to: lodging, where plants break in the wind and the seeds fall to the ground, germinating in the following spring; shelling, which occurs mainly in oilseed crops, when the mature plant becomes brittle and the movement by harvesting equipment causes some of the pods to shatter prior to being harvested, falling to the ground and germinating the next spring; a wet harvest, which can cause some low-lying crop to be left in the field to germinate the next season, or wind that can blow swaths apart leaving portions unharvested; animals or fowl feeding on crops in the autumn and scattering heads of grain from a swath that are not harvested; and harvesting equipment for grain and oilseeds that, no matter how well set, puts over a small percentage of seed that germinates in spring.

Many of these factors can combine together, with the result that a large number of seeds often remain in the field. Gulden *et al.* (2003) have estimated that as many as 3,000 canola seeds per square metre can lie in the soil following harvesting. When planting occurs the next spring, these seeds are present to germinate and create the problem of controlling volunteers. When this happens with GM volunteer canola, for example, spraying with 2,4-D can control the volunteer canola. However this chemical application means an additional cost of C$1.50-C$2.00/acre to the producer, while for organic producers this is not an option. The introduction of HT wheat is expected to make control of volunteers more difficult as 2,4-D does not control volunteer wheat, with the result that producers will have to use a more expensive chemical when controlling HT wheat volunteers. Officials with Monsanto have suggested that the most cost effective method to control volunteer GM wheat will be to tank-mix and apply Roundup® and Assure®, at an estimated cost of C$6.19/acre. This method of volunteer GM wheat control is more than triple the cost of volunteer GM canola control. Officials with Aventis (now Bayer) suggest that all maize chemicals, with the exception of Liberty™, can be used to control volunteer maize. The application rates and costs of these chemicals vary. A recent study prepared for the Canola Council of Canada (2001), which surveyed 650 western Canadian canola growers, showed that 16% of the growers said that managing volunteers with HT canola was easier than with conventional canola varieties, while 23% said that it was more difficult to manage.

Cross-pollination is an issue that has great importance to commercial agriculture, yet in some crops minimal literature is available on this subject (Table 7.1). For example, existing research conducted on the drift of wheat pollen was done in Saskatchewan in the 1930s. This has resulted in a research gap of over 60 years and studies that are presently underway are challenging the standards that are presently in place to prevent wheat cross-pollination (Hucl, 1996 and Hucl and Matus-Cádiz, 2001). In maize, Losey's (1999) study was the first to examine maize pollen drift in over 25 years. This study spawned numerous other studies that were conducted in 1999 and 2000, many of which concluded that 90% of maize pollen was deposited within 5m of the edge of the maize field. Canola pollen, however, is dispersed over a much wider range. In one instance canola pollen has been traced to a distance of 25km (B. Kennedy, 1999).

Table 7.1: Impacts from GM crops

Issue	Canola	Wheat	Maize
Potential to out-cross	Yes	Yes (limited)	Yes (limited)
Detected distance of pollen drift	10-25 km	200m for some varieties	100m
Chemical control of volunteers	Yes, C$1.50-2.00/acre	Yes C$6.19/acre	Yes, varying rates
Required isolation distance between plots for seed	100m for like varieties, 10 to 800m for other canola crops	3m for other crops and 1m for same variety	Ranges from 15m to 200m depending on plot size

Source: Canadian Seed Growers Association; Staniland *et al.*, 2000; and authors

Canola, for example, is frequently an open pollinating crop, which means that HT varieties can cross-pollinate with each other, with conventional varieties and with weedy relatives. This has resulted in cross-pollinated hybrids that are resistant to more than one chemical. In 1999, the first triple-resistant canola was discovered in Alberta (Western Producer, 2000). These plants were subjected to chemical and DNA testing and were found to be resistant to Roundup®, Liberty™ and Pursuit™. While many in the canola industry were predicting that this would eventually occur, this triple-resistant hybrid was created in just two years. While resistant to new stronger chemicals, the hybrid variety is still susceptible to 2,4-D and can be controlled. The concern among many producers is that other crops, such as wheat, may already have developed resistance to some chemicals, making it more difficult to control cross-pollinated weeds or volunteers, with the result that efforts may be extremely expensive or all but impossible.

Canola may represent one extreme as it is open pollinating. Recent research from France has examined the potential for genes from rapeseed to flow into wild mustard, hoary mustard and wild radish (Chèvre *et al.*, 2000). This study found that on average, the rate of out-crossing was 0.18% for wild mustard, 1.9% for hoary mustard and 23.8% for wild radish. Collaborative research between scientists in Canada and France (Lefol *et al.*, 1996a) has shown that cross-pollination between canola and wild mustard is virtually non-existent. This study examined 2.9 million wild mustard seeds and concluded that no hybrids were found and that actual cross-fertilisation appeared to be below one per million (*ibid.*). A study on the possible hybridisation between canola and hoary mustard (Lefol *et al.*, 1996b) found that while it was technically possible, the hoary mustard seed had to imported from France to enable the study to take place in Canada, as hoary mustard is unable to survive the winter season on the Canadian prairies. While wild radish is a weed in the Maritimes of Canada (with only one sighting in Alberta), given limited canola production there, the potential for gene escape into wild radish was judged to be remote at best.

Given this level of biological uncertainty, volunteers and cross-pollinated varieties or weeds are inevitably going to be co-mingled in the commodity food system. Regardless of the market, consumer surveys continue to show significant and rising preferences for organic and non-GM products. At a minimum, consumers simply want to know what they are eating, be it organic, GM-free or GM foods. To foster consumers' trust in products labelled non-GM and organic, control of GM cross-pollination and volunteer GM seed will be essential.

There are a number of examples where co-mingling has imposed significant costs on an industry. Perhaps the best-known case is the one concerning Aventis' StarLink™ maize. That variety was approved for use in the US as an animal feed and was required to be produced in segregated areas, surrounded by a buffer crop, which was also supposed to be marketed as feed. As outlined in Chapter 2, the GM trait in the feed maize was found in the human food chain, contaminating an estimated 10% of all foods containing maize meal. The costs of containing and removing the offending variety have been substantial. Aventis has settled this case, with US$110 million going to compensate producers and pay for the logistics of withdrawing the variety, while many food manufacturers, such as Taco Bell, have had to recall whole product lines that have been contaminated.

While the StarLink™ contamination was an extreme case, it does not seem to have destroyed public confidence in the entire product line. There is, however a significant possibility that contaminations could jeopardise entire product lines. Some have argued that North America can no longer sell organic crops to the EU while others disagree.

As discussed in Chapter 2, it is not clear yet who bears ultimate liability for cross contaminations or co-mingling. The StarLink™ incident spawned numerous lawsuits by producers, producer organisations and a number of states (e.g. Missouri) against Aventis in an attempt to seek compensation for depressed maize prices that they claim resulted from lost foreign sales. Similarly, a pending counter-suit in Canada by Mr. Percy Schmeiser against Monsanto argues that because Monsanto owns the intellectual property in Roundup Ready canola, it also should be liable for any lost canola sales due to contamination. The Saskatchewan Organic Directorate has also filed legal action against Monsanto and Aventis claiming the destruction of the organic canola export market.

While co-mingling and adverse market responses are important, the problems of managing gene flow also have significant potential to lower the diffusion and adoption of new technologies, and hence lower the commercial and social benefits of the investments. The problem is that conventional gene controls are not effective enough.

An incident in Europe highlights the challenges facing governments and industry. In the spring of 2000 it was announced that the EU found a breeders' lot of canola seed imported by Advanta that contained 0.4% unapproved GM traits. Advanta quickly determined that the unexpected presence of GM canola was caused by gene flow from GM foundation seeds that had been planted in a neighbouring field. Canadian seed growers had followed isolation rules but the genes still moved into the conventional foundation seed. While the total acreage in most countries was insignificant (Sweden and Germany had 300ha and France had 600ha), the outrage expressed by environmental groups, the media and some government officials surprised the Canadian canola industry. While many in the Canadian canola industry termed the EU response an 'over-reaction' and felt that they were acting with 'hysteria', this incident highlights the need for a technology that can prevent the recurrence of similar incidents. The European countries faced a cost in dealing with this problem as France ' ordered all 600ha to be ploughed down and Sweden allowed the seed to be harvested but prohibited the canola from entering the domestic or European market.

Conventional containment regulations can also make adoption of new crops prohibitive. Many producers only adopt new crop varieties after watching a neighbour have success with that variety. A common practice for producers in Western Canada is to seed 80 acres of a new variety as a test before fully adopting the variety. When Monsanto and AgrEvo introduced their HT canola varieties, they did so with 80-acre production contracts as they believed this was the most effective method for producers to evaluate the new technology. The increased use of buffer zones to control cross-pollination has the potential to drastically reduce the adoption rate of new technology crops. If the StarLink™ buffer zone of 660 feet is used as a base, this entirely removes the option of 80-acre production contracts as the buffer zone consumes the entire 80 acres. Moving to 160-acre production contracts is still very restrictive, as 76% of the land would be consumed in the buffer zone. Producers would be required to plant 40 acres in the centre of a quarter section, a sub-optimal evaluation size, and plant 120 acres to a crop that provides sub-optimal rent.

Finally, ineffective IPRs in many countries reduce their attractiveness as markets for new technologies, causing them to lag in the adoption of new traits and varieties. In the canola sector, for example, few companies would choose to export new cultivars to major growing regions in China or India because of the lack of effective IPR protection. As a result, about half of the producers of rapeseed/canola in the world are unable to access the latest technologies, which is one of the contributing factors to lower yields in those areas. India, for instance, posts average canola crop yields almost 40% below Canada's while China, in spite of significant subsidies for irrigation and fertiliser use, still posts yields about 3% below Canada's. Finding a more effective IPR mechanism that is not dependent on institutions that are often very weak in many of these countries might improve diffusion of new cultivars and technologies. If developing country yields were to rise even 5% due to new varieties, total canola production there would rise by about one million tonnes, worth approximately US$225 million to those producers and their markets.

In brief, plants and people cannot be trusted to do what markets require. As a result, genes will move, creating co-mingled traits in the food system and liabilities in the transfer of technologies between markets.

Organisational and Biological Control of Liabilities

Fundamentally, either institutions or biological controls or a combination of the two can manage risks. Institutionally, the public sector evaluates new GM crops for safety considerations, examining the new products against known products to determine whether they involve any new risks related to human consumption, the environment and livestock (if used as feeds). If the new variety is determined to be substantially equivalent, then it will usually be approved for release. Most regulators also have some ability to examine risks once the products enter the market and may intervene if an unexpected risk is detected. While some of these products might be only conditionally released (e.g. for production in a specified area or under conditions of isolation from food crops), most will be released without condition. In both cases, the private sector is generally responsible for managing the liability of new GM products once they enter the market. They use a combination of contracts, testing and auditing to ensure conformity. While these mechanisms are very important, they cannot manage all the liabilities— genes are likely to travel. Regardless of how effective regulations or contracts are, some actors will either deliberately or inadvertently misuse new technologies, creating potential new liabilities. More importantly, however, many plant species are promiscuous sexually, creating natural gene flows. In short, we need to look for biological control mechanisms to manage many of the resulting liabilities of these new crops.

In flowering plants, pollen is responsible for delivery of male gametes to the female reproductive organ (the carpel) of the same or another plant, respectively resulting in self- or cross-pollination (Sawhney, 2001). Thus, pollen development and its transport are crucial for successful sexual reproduction in angiosperms, and for subsequent fruit and seed development. Pollens are produced in large numbers and are transferred to the carpels by vehicles such as wind, animals and insects. Pollination as a process allows for a limited fertilisation of plant ova across the species spectrum. In order for pollination to be successful, physical attributes of pollen must be genetically competent to endure the physical carriers and fulfil their intended function. In regions of the world with harsh winter climates, overwintering of sugar beets, for example, aids plants to survive the winter, flower in spring and become a source of pollen dispersal. If sugar beet plants with genetically modified traits are planted, Pohl-Ort *et al.* (2000) indicate that pollen dispersal with wild or weed beet varieties could occur after over wintering. Interference with these genetic and environmental attributes could become the choice of scientific options for control of the GM gene flow.

The control of pollination would involve the option of genetic manipulation of its development or function. Pollen biotechnology, which Sawhney (2001, p. 120) defines as "manipulation of pollen development and/or function with the objectives(s) of increased production and improvement of crops and other pollen products," may become the main control mechanism in GM gene flow. There are several loci for genetic manipulation and interference with steps of pollen development. Obvious examples of interference are: pre-meiotic and meiotic events in the microsporogenous tissue; the development of microspores and pollen in the anther; pollen maturation and pollen release; pollen dispersal; the attachment of pollen to the stigma of the carpel; pollen germination and tube growth; and the release of sperm cells into the female gametophytes in the ovule of an ovary. Genetic blocking of genetic functions governing these events can lead to absence of seed and fruit development. As an applied example, plant breeders and molecular geneticists have made use of male sterility traits to place a preferred definition for particular and advantageous use (e.g. those associated with fruit processing or gustatory values). However, in the past these definitions were not necessarily made for curtailing the propagation of the crops outside their intended area or restricting growers' access to the seeds.

Given that pollination is simply the means for distributing genetic material, pollens with an incomplete set of genetic material would potentially impede pollination. The 1990s brought new efforts to ensure sterility, not for processing or gustatory value, but for commercial IPR protection. Efforts are underway to limit the diffusion of transgenes through genetic use restriction technologies (GURTs) to turn off reproduction for either transgenic varieties or traits. Some dubbed this approach the 'terminator gene'. Patent data suggest that three competing systems are being worked upon. Two examples are found in the patents of Tomes (1997) and Odell (1991). Tomes' patent (Patent WO 97/40179) combines two independent biochemical traits (i.e. synthesis of tryptophan) and indole acetic acid (IAA). Indole acetic acid is an auxin whose over-expression leads to seed abortion. The Odell patent (Patent WO 91/09957) describes how *in-vitro* rDNA techniques generate two genes (Cre-loxP) from bacterial virus to produce a number of DNA site-specific cytotoxicities and inactivation of seeds. The expressions of these genes result in the second generation of seeds of a crop being infertile, rendering such seeds useless for repeat seeding. The latter system is based on orchestrating a set of preferred genes (RIP, LEA, CRE/LOX, Tn10 tet) that are organised in a cassette of genes that, upon a particular environmental stimulus (e.g. presence of a chemical, temperature-osmotic change or shock), will disallow production of viable seeds. Hence, these are described 'terminator' genes.

The use of sterile seeds *per se* is widely practised and has not raised objections while the GURTS technology is being criticised because it could deny poor farmers the option of saving seeds for future use. However, the crudest form of this technology was known and documented long before notions of inheritance and genetics were known. Seedless grapes were known to have existed from writings of ancient Egyptians and Greeks (c. 3000 BC). Since the 1930s seedless edible crop varieties produced by traditional plant breeding methods have been produced (e.g. seedless grapes in 1936 and watermelons in 1951) and possess certain advantages (Table 7.2). The prerequisite knowledge of flower pollination, fertilisation, fruit development, genetics and rDNA-based technologies has now enabled scientists to apply this technology to new crops. To generate seedless fruits, strategies of genetic interference at the level of post fertilisation and seed development, parthenocarpy or a combination of both can be contemplated. In certain seedless varieties (e.g. tomatoes, bananas and pineapples) the trait exists because the ovaries of such plants are able to develop via a process called parthenocarpy, which does not depend on fertilisation. Additionally, prevention of seed production could be achieved by interfering with the process of pollination.

Table 7.2: Advantages of seedless fruit and vegetable crops

	Fruit quality	Shelf life	Taste	Processing	Production	IPRs
Citrus	More tissue	NA	Better	Juice	NA	Yes
Cucumber	Crunchier	NA	Better	Pickles	NA	Yes
Grapes	More tissue for raisins	NA	Better	Juice	NA	Yes
Tomato	More tissue	NA	Better	Juice and ketchup	NA	Yes
Watermelon	More tissue	NA	Better	Juice	NA	Yes

NA = no advantage over seed variety

Source: Authors

Ultimately, it may well be a question of choice—that is, whether the 'terminator' technology is the only option. There is the possibility that many of the concerns about genetically modified crops could be overcome by further advancements in science (Daniell, 1999). A number of biological options exist, depending on the crop and its attributes. Both traditional and molecular genetic methods already provide mechanisms to create hybrids, while working at a more refined molecular level offers the potential to control GM traits. Recently there has been an effort to reduce the risk of biotechnology crops by engineering foreign genes via the chloroplast instead of the nuclear genome. Such recombinants would only express the new traits in selected parts of the plant, rather than in the whole plant. Hence, any pollen drift would not include the transgenes. These and other options offer some promise.

'Terminator' technology, which has come to symbolise all the possible scientific options, is the end result of an evolved 'normal science' process (see Chapter 3). It could be argued that because Monsanto has withdrawn its use, efforts to develop that concept and other technologies will cease. This may, however, be a classic case of paradigm shift and start of another wave of 'normal science' that targets the new tools of genetics and biotechnology to management and control of GM gene flow to non-GM plants in the first instance and to bolster IPRs in the second.

Biological control of liabilities, either through contemporary technologies described above or those yet to be devised, is the science side of the story. The human, institutional element is the complementary other side. Ultimately, these two parts must fit together in a discussion of the relative costs (risks) and benefits of alternative options. As noted above, the costs of not managing the liabilities are potentially very high, ranging up to the US$110 million cost of the StarLink^{TM} failure. Similarly, control mechanisms are not cheap. Often incomplete institutional approaches can cost in the millions for those technologies that are widely dispersed, with both high fixed and variable costs. One potential advantage of the GURTs biological control mechanism is that while it is costly to develop, the marginal cost may be as low as US$250,000 per new variety released (Visser *et al.*, 2001), which would add only about 10% to the cost of a new commercial variety. Given that many firms report they lose at least 10% of their returns due to incomplete property rights, this option may be significantly more effective than other approaches.

Conclusions

Regulators will be faced with two options regarding new output-trait GM crops: they can reject them outright or they can impose detailed production and market segregation regulations. The outright rejection of new biotech crop varieties may be excessive given the level of risk and potential benefits. Numerous benefits have already been suggested for new generation GM crops such as Golden Rice™, tobacco that fights cancer, tomatoes that reduce heart disease and cholesterol-reducing grains. If new risk management measures are excessive, they could certainly jeopardise future R&D investments in those areas.

In this chapter we argue that GURTs types of technology could provide some advantages. First, they could act as built-in safety mechanisms to prevent the escape of potentially harmful traits (e.g. HT) from new GM crops. Second, they could prevent pirate growers from exploiting GM seeds and compounding risks. Third, they could reduce product liabilities assigned to the seed growers by preventing contamination with non-transgenic crops.

There would appear to be a three-pronged approach that could realise the benefits of new crops while at the same time minimising risks. On the institutional side, governments can and should improve the regulatory oversight of second and third generation GM crops, possibly aggressively using refugia, contract registration, regional regulation and mandatory crop rotations and audits. Meanwhile, industry must take its responsibilities more seriously. The introduction of first generation GM products was directed at getting producers to adopt the technology, and many of the firms do not appear to have a strong appreciation for the importance of managing the technology and containing it. Moreover, the launch of these products went largely unnoticed by consumers and much of the industry. This approach cannot and will not work for second and third generation products. Many of the potential liabilities of GM crops are only partially manageable by public and private institutions. Institutional costs to manage risks are high, while the cost of failure is even higher. Ultimately, inability to manage the risks and control the liabilities may sink the technology.

As has been demonstrated, the lack of control mechanisms for GM pollen and seed is presently affecting producers and exporters of not only crops and oilseeds, but other products as well. To continue in this direction where no control mechanisms exist will only result in higher costs for seed development companies and producers. With domestic subsidies on the decline, affected producers may turn to litigation as a means of recouping lost revenues (see Chapter 2). Regulators and industry officials need to examine what the market impacts would be from commercially releasing a control mechanism for GM crops versus leaving the situation as it is, with an expected rise in litigation costs. While the initial cost of introducing a control mechanism may be high, the long-term benefits of such a technology may justify commercialisation.

In short, even with the best institutions, some risks will remain. Hence, biological control mechanisms need to be considered.

Chapter Eight:

Supply Chain Responses to Liability

Introduction

The global agri-food system faces one of its greatest challenges in more than a century as it seeks to adjust to changing consumer demands. Over the past 100 years, the agri-food sector has invested heavily in regulations, processes, germplasm, breeding stock and market structures to homogenise food products to facilitate the production and trade of uniform products at affordable prices. Bread, beef, wine and many other products became much more homogeneous within and between markets. In recent years, however, many consumers, especially those in higher income countries, are demanding much greater differentiation in these basic products. This is at times being aided and abetted by producers and governments using this 'opportunity' to erect barriers to international trade in competing products.

Consumers' search for variety now extends beyond the physical attributes of food to include how their food was produced (e.g., by region, by certain producers and using certain technologies). The food trade has responded recently with rapid differentiation in products. Organic production is growing, firms are marketing ethically, environmentally and socially responsible products (e.g. free-range chicken in Britain, green label foods in Canada and producer-owned coffee from Latin America), GM-free foods are entering the markets in North America and Europe and producers around the world are looking at using Appellation Controle to differentiate their produce.

While it is relatively easy for producers to label their products as meeting certain specifications, fraud laws, international agreements on food labels and many consumers require assurances that such foods actually have the advertised attributes. Spot markets are unable to provide those assurances. As a result, governments and companies are looking for ways to assure the quality of food. As discussed further in Chapter 9, various segregation, identity-preserved production and marketing (IPPM) and traceability systems are evolving to bridge the gap between differentiated consumers' wants and the traditional agri-food production and marketing systems that deliver homogeneous commodities.

The Pressures for Change

A number of events have converged in recent years to increase the need for identity preservation in the agri-food system. In the first instance, consumer preferences have been changing. Higher per capita incomes, increased concerns about the safety and appropriateness of the food production system and more international migration have helped to increase the diversity of demands for dietary and non-dietary features in food. This is especially true in the OECD countries that are both the source and destination of many traded foodstuffs.

Second, technology has advanced, creating both a greater range of processing techniques and a wider variety of foods with new attributes. Biotechnology, in particular, is delivering a range of animal and plant varieties with new input and output traits. In one very important area, however, the technology lags. While it is now possible to produce an extensive range of new foods and to distribute them widely, it is often very costly or impossible to test these foods to determine what attributes they have. Many new crop and livestock products exhibit both experiential and credence elements, involving either input traits that entail some public interest or output traits that only have value if identity-preserved. The inability to search for the attributes necessitates more managed markets with credible quality signals. On the plus side, advances in transportation, communications and information technologies make it technically feasible to both differentiate and identity preserve a wider array of products, many with attributes that go beyond physical characteristics.

Third, agri-food markets are becoming truly global. Large agri-food companies, operating in multiple markets around the world, have both created pressures for, and benefited from, recent international trade agreements that begin to liberalise trade in agricultural products. The new WTO and its ancillary Agreements on Sanitary and Phytosanitary Measures and Technical Barriers to Trade, build upon the sectoral and product agreements managed through the International Plant Protection Convention, International Epizootics Organisation and the Codex Alimentarius. Processed foods are now often comprised of food products originating from a large number of countries. Processing is multi-staged and undertaken along supply chains that cross numerous international borders.

Fourth, domestic regulators are being forced to respond to domestic citizen and consumer concerns about new foods and new technologies, in advance of any international consensus on how to handle them. As a result, regulation of food quality and safety is rising around the world, at times creating new hurdles for international trade in food. This happens most often when countries are faced with new risks or new technologies. Bovine spongiform encephalopathy (BSE), genetically modified organisms, organic foods and ethical or religious concerns about animal husbandry or processing methods have led to significant changes in many domestic regulatory systems (see Chapter 5 on domestic regulation). Some countries have opted for a more cautious approach to food risks, choosing to delay introduction of new products or to ban certain practices, while others have more aggressively adopted and exploited new technologies. As a result, some countries do not allow some production and processing methods, and ban trade in products that have used those methods. This poses significant challenges for exporters who might have a product that may meet the importer's domestic standards but is difficult to verify. Somehow they need to assure the importer that their trade does not infringe on domestic authorities. Testing is often either prohibitively expensive or not possible. At the same time, most of the major food exporting countries—e.g. Australia, Canada, New Zealand and the US— have indicated to the OECD that they are aggressively moving towards more industry involvement in food quality and safety, through industry-based standards systems such as GMP or HACCP systems. While this may reduce domestic regulatory costs, the standards underlying these industry-based systems are not yet accepted universally.

These converging events have tended to cause a greater fragmentation in world food markets. There are now GM and GM-free markets (North American vs. EU) while trade in some products from specific markets has been banned (e.g. UK beef), even when some of the produce does not pose a discernible risk. Trade has slowed in some product lines and between some groups of countries with differing perspectives on quality. Recently, firms have begun to develop new quality-assured supply chains to bridge the gaps between those markets. Organic producers in North America and around the world are developing standards and procedures to ensure their products are credible. Maize and soybean producers have tried to develop and deliver quality assured GM-free grains to Europe and Japan (Lin, 2002). Beef producers in the UK, US, Australia and Canada are developing quality assurance and traceback systems (Spriggs and Isaac, 2001). Canola producers and processors in Canada have developed IPPM systems to manage more than 39 new varieties with differentiated traits (Smyth and Phillips, 2003).

Historically, crop production from different fields and different farms has been pooled for transportation purposes. Currently, there is a move to increasingly differentiated production. Differentiation was initially used by the crops industry to distinguish between grades of grains and oilseeds. The industry has now progressed to the point that it uses differentiation for organic production, speciality contracts for crops with particular attributes and buyer-specific purchase requirements. Plant breeding technology over the years has created a large number of varieties with input and output traits that require differentiation to maintain their value. Biotechnology has increased the number of varieties requiring differentiation, which has led to the rise in use of IPPM systems. As end users continue to refine their purchase requirements, the number of IPPM systems will increase and the volume these systems are capable of handling will rise. Initial grade differentiation has been extrapolated to accommodate the increase in grain varieties requiring IPPM systems. As consumer demand for specific food attributes grows in importance, IPPM systems are expected to become commonplace in agriculture.

In the absence of an internationally agreed set of rules for the regulation of trade in biologically-based products, nations will have to establish domestic trade regimes that address the concerns of consumers. While some countries may feel that consumer demands for protection are sufficiently small to be safely ignored in trade policy (and thus will be willing to allow unregulated imports), others will not. The latter are faced with a choice between implementing an import embargo or a labelling regime. The problem for process attributes such as growth hormones or transgenic genes is information asymmetry along the supply chain. Only the producers of inputs embodying the new process attribute and farmers using these have full information regarding whether the product they sell was produced using the new product attribute. In the absence of an effective IPPM system, the possible co-mingling of products causes information to become progressively more incomplete as the product moves downstream along the supply chain and into international distribution channels. Thus, final consumers in importing countries will not be able to determine if the imported product was produced using the new process attribute.

In the absence of an IPPM system, only a single blended market or pooling equilibrium can exist. Adverse selection characterises such markets and they will become dominated by an inefficient proportion of low quality product, or what Akerlof (1970) calls 'lemons' (Plunkett and Gaisford, 2000). A fully effective IPPM system could lead to separate markets for products produced using existing process attributes and new process attributes. The different tolerances for GM ingredients in the differentiated markets creates the potential for large commercial and socio-economic liabilities arising from incomplete performance.

This chapter examines an array of IPPM systems, using examples from the Canadian canola market since 1995. A number of national regulatory systems have evolved to manage public expectations and concerns about these products.

The Conceptual Framework

Ultimately, differentiation is about quality. Quality is a multifaceted aspect for any product. Neo-classical economic theory suggests that two key elements are vital to the creation of quality: consumer tastes and preferences and producer efforts to develop consistent, safe, affordable, attractive fare that meets consumer demand. Theory suggests that most, if not all, of these elements can and should be produced within minimally regulated markets.

Increasingly, however, the literature is pointing to variables of trust and confidence in the creation and operation of markets (e.g., Fukuyama, 1995; Stiglitz, 1999). As pointed out previously, markets for many products are not able to create, by themselves, the conditions of trust that generate the socially optimal quantities of goods and services produced and consumed. The divergence between the optimal and actual outcomes could generate socio-economic liabilities. Hence, there is a much more explicit role for public and private regulation in markets than neoclassical theory generally suggests. This is particularly true for GM agri-food products, where perceived risks and public uncertainties abound.

In a great many instances in the market place, a simple arms-length exchange of goods and services at an agreed upon price is a low-cost transaction that provides the correct incentives for buyers and sellers. Williamson (1979) notes that where transaction costs are negligible, the organisation of economic activity is irrelevant, as any advantages of one mode of organisation over another will be eliminated by costless contracting. Williamson (1979) and others, however, suggest that three principal dimensions of transactions may create conditions that make alternative market structures more efficient. First, uncertainty due to opportunistic behaviour can increase the cost of using spot markets. Second, the frequency with which a transaction recurs increases the opportunity for specialised mechanisms. Third, high asset specificity (i.e. an asset with a low opportunity cost in its best alternative use when the original transaction is terminated) increases the risks of using spot markets.

As discussed in Chapter 4, the attributes of many of the new crop varieties create differing levels of risk in the market. Many of the new GM crop varieties exhibit both experiential and credence elements, involving either input traits that entail some public concern or output traits that only have value if differentiated. The inability to search for the attributes necessitates more managed markets. The combination of significant potential for opportunistic activity (as some producers may wish to place low value crops into high value markets) and high asset specificity for many of the proprietary traits or for the output trait has created the incentive for a wide array of managed supply chains. In the first instance, where there are no novel output traits or concerns related to the technology (e.g. no credence factors), spot markets are probably the most efficient market structure. When GM technologies are used, credence factors arise for some consumers, creating opportunities to differentiate and increase social welfare. Given the difficulty of searching for the differentiable traits, such differentiation will inevitably create specific assets and ample opportunities for cheating, necessitating a more integrated, managed supply chain. Finally, all output traits, whether conventionally or transgenically inserted, will generate significant challenges that will necessitate both public and private action. The two integrated chains will differ depending on the potential for real, measurable health and safety impacts. Where there are known or anticipated impacts, the state will inevitably reserve a role in managing the system while markets will probably be left to manage those systems where the concerns relate to preferences.

Jacquemin (1987, p. 138) states that "hierarchies, federations of firms, and markets compete with each other to provide co-ordination, allocation and monitoring. It is only when one organisational form promises a higher net return for specific activities than alternative institutional arrangements that it will survive in the long run." This return is net of liability costs. Just which ones will survive in the biotechnology supply chain is yet to be determined.

The Economics of Current IPPM Systems

The ultimate question facing the industry is whether it makes any economic or commercial sense to develop an IPPM system to serve the GM-free or other differentiated food markets. Ultimately, the answer depends on the cost of IPPM systems, the sensitivities in the systems, the risks and their distribution, strategic options available and the potential liabilities that can arise.

The research on IPPM systems is limited, but growing (Table 8.1). The widely recognised starting point in this literature is the study done by Buckwell, *et al.* (1999) that argues that IPPM would only be a 5%-15% cost increase above farm gate prices. This widely circulated, but somewhat thin, study was an examination of several IPPM systems that operated in the US, Brazil, Europe and Canada. The only identifiable study prior to Buckwell *et al.*'s was a short EuropaBio (1997) study that concluded that IPPM could raise costs by 140-180%. Several separate research projects started in 1999 and arrived at varied conclusions. The Economic Research Services, USDA, released two studies that examined IPPM. Golder and Leung (2000) looked at labelling for genetically modified (GM) products and found that there was a premium for non-GM maize and soybeans of 2-6%. Lin (2002) looked at differentiating non-GM maize and soybeans and found that costs for differentiation ranged from US$0.18-0.54 per bushel or US$12.47-28.54/tonne respectively. These estimates are based solely on the IPPM costs incurred by the primary elevator system and additional transportation costs associated with delivering product to export terminals. The estimate is comprised of extra costs for storage, handling, risk management, transportation, testing and marketing. Similar research by Bender *et al.* (1999) on non-GM maize and soybeans found differentiation costs of US$0.18-0.33 per bushel, but warn that caution must be used when interpreting the cost of handling speciality crops due to the very small number of firms involved and the low commodity volume.

Using a theoretical model, Maltsbarger and Kalaitzandonakes (2000) found that IPPM costs ranged from US$0.17-0.27 per bushel in large inland elevators. A group of researchers from the Iowa State University released an economic analysis of differentiation (Miranowski *et al.,* 1999) that offers several academic perspectives. A subsequent study from Iowa State looks at the level of premiums paid to producers for output trait speciality maize varieties. Ginder *et al.* (2000) state that 64% of those surveyed estimated that the cost to handle speciality maize ranged from US$0.05-0.25 per bushel. The University of California at Davis released a study on IPPM that focuses on the rice market in California. Herrman *et al.* (1999) estimates the costs in the US of differentiating wheat into two or three individual products. The transportation sector has had two specific studies. The first, by Vido *et al.* (2000), is a study from the University of Manitoba that looks at whether separate supply chains will need to be established for GM products. The second, a study from North Dakota State University by Reichert and Vachal (2000), examines rail shipping costs for IPPM products. Finally, Gosnell (2001) examines how the Canadian wheat industry could cope with the introduction of GM wheat. His research concludes that the differentiation between GM wheat and non-GM wheat is most economically efficient if it is done within the receiving grain terminal.

Finally, two specific studies relate to canola. Smyth and Phillips (2002) examined IPPM systems that have operated and are operating, both within Canada and the US and documented that IPPM systems cost 15-20% above the cost of handling conventional commodities. It is important to note that most of these costs are only calculated to the point of reaching the processor and it would be reasonable to expect that the final cost of an IPPM system would raise this cost. Sparks Companies Inc. (2001) in their review of IPPM systems has estimated that non-GM canola would require a premium of C$26-31/tonne to be delivered to Vancouver ports for export. In the same study they estimated that non-GM soybeans would require a C$36-41/tonne premium over GM varieties to be supplied at port. The majority of these costs are associated with the producer segment of the supply chain. For canola, they estimated that producers would require a premium of C$16/tonne to undertake IPPM. For soybeans they estimated that producers require to be compensated from C$22-24/tonne to begin implementing an IPPM system.

Table 8.1: Summary of IPPM system costs

Commodity (terms)	Year	IPPM Cost
Non-GM Canola (FOB Vancouver vessel; minimum) [a]	Est.	C$25.90-$30.65
Non-GM Soybeans (FOB export position; minimum) [a]	Est.	US$35.53-$40.92
Non-GM Maize (primary elevator to export elevator) [b]	Est.	US$12.47
Non-GM Soybeans (primary elevator to export elevator) [b]	Est.	US$28.54
Food Maize (FOB inland elevator) [c]	1998	US$43.22
High Oil Maize (FOB inland elevator) [c]	1998	US$14.01
Food Soybeans (FOB inland elevator) [c]	1998	US$91.58
STS Soybeans (FOB inland elevator) [c]	1998	US$17.99
GM canola (crushed domestically) [d]	1996	C$33 - $41
Soybeans (container, in store Japan, no producer costs or testing) [e]	2000	US$27.72

Sources: [a]Sparks Companies Inc., 2001, [b]Lin, 2002, [c]Bender *et al.*, 1999, [d]Smyth and Phillips, 2002, [e]Reichert and Vachal, 2000.

One key concern about the recent estimates is that many of the IPPM systems studied are only used for handling niche market products. The low volume handled by these niche IPPM systems does not create confidence in the cost estimates derived from the systems. In Canada, the largest IPPM system operating, the Warburtons wheat system, operates at a volume of 200,000-250,000 tonnes annually. In the US, Bender *et al.* (1999) identified that some differentiation systems for varieties of speciality maize approach 50,000 tonnes. Most of this speciality maize is not handled through strict IPPM systems; rather, it is essentially binned at country elevators separately from other varieties of maize with minimal concern about co-mingling.

In conclusion, one could argue that it is too early to say what the real costs of IPPM systems will be. All of the systems analysed involved small volumes and purpose built systems for discrete markets. If IPPM expands, risks will multiply and costs could increase. Alternatively, higher volumes and a greater array of systems could bring some economies of scale, which would lower the cost. It is impossible to say with the data available which factor will dominate.

Evolving Crop Production and Marketing Systems

A more detailed examination of the Canadian canola industry provides some insight into the problems facing IPPM systems in the global agri-food system. The introduction in 1995 of two HT varieties of canola precipitated the first IPPM system for a GM crop in Canada. If the new varieties introduced in Canada but pending approval in Japan or the EU were allowed to co-mingle with approved varieties, the resulting shipments to export markets would be jeopardised. Table 8.2 presents the distribution of Canadian production by consumer market. Given this market structure and the accelerating flow of new varieties, Canadian firms handling canola had three options. The firms could commercialise the new varieties, co-mingle production and lose access to the EU and Japanese markets, which in 1994-5 imported 42% of the Canadian output. The firms could withhold their new varieties until they were approved in all key markets. As pointed out in previous chapters, Heller (1995) estimates that a regulatory delay of even one year decreases the rate of return on a new product by 2.8% while a two-year delay decreases it by 5.2%. Alternatively, Canadian producers and exporters could accept responsibility to differentiate GM from traditional canola and develop a system to provide quality assurance of delivery to the key export markets, which would involve Canadian export companies developing new systems of identity preservation in production and marketing.

As the Government of Canada does not have the legal mandate to govern the exporting of GM canola, the industry chose to take the initiative and develop the export rules needed to assure continued access to foreign markets. The research/seed companies and grain companies shared this task. This group developed a series of vertically-managed strategic alliances through proprietary supply chains and horizontally through the Canola Council of Canada to manage the flow of GM product in order to allay concerns in the Japanese and EU markets.

Table 8.2: Canadian canola production and export destinations

	Canadian Production	% total production consumed domestically	% of production flowing to key export markets			
	(000t)		Total	Japan	EU	US
1992-3	3,872	22%	78%	42%	8%	21%
1993-4	5,480	15%	85%	38%	19%	21%
1994-5	7,233 .	26%	74%	25%	17%	19%
1995-6	6,436	30%	70%	28%	6%	24%
1996-7	5,056	12%	88%	42%	4%	26%
1997-8	6,266	15%	85%	34%	1%	31%
1998-9	7,588	21%	79%	31%	0%	24%
1999-0	8798	33%*	67%	31%	0%	23%
2000-1	7088	4%**	96%	27%	0%	25%
2001-2	5062	12%	88%	31%	0%	27%

* Due to record production in this crop year, domestic carry-over was 1.5-2 times normal

**Domestic consumption still low due to record production in previous year

Source: exports are the sum of seed, oil and meal trade; retrieved from the world-wide web November 20, 2003 from http://www.canola-council.org/stats

Over the intervening years, five identifiable types of supply chains have been developed and have operated in the canola industry in Canada (Table 8.3): for input traits (spot markets and voluntary general grower contracts for both non-GM and GM input traits) and for output traits (voluntary grower contracts for non-GM output traits and mandated IPPM systems for both non-GM and GM industrial output traits). The data for this analysis were gathered between 1997 and 2001 from numerous public and private sources. Public variety data were gathered and analysed. A small survey was sent to a select number of canola seed development firms in 1999 to determine the importance of intellectual property rights, quality assurance and identity preservation in supply chain development. Several interviews were conducted to gather the remainder of the information needed for this analysis. Industry participation was crucial for this case study as several private seed development firms provided information that was not otherwise available.

Table 8.3: Illustrative examples of types of supply chains in the Canadian canola industry, 1990-2001

	Regulatory base	Type/source	Illustrative variety(s)	Special trait	Dates	Marketing mechanism	Seed developer	Grain merchant	Acres
Non-GM input traits	Variety standards in Seeds Act and trademark for canola	1	AC-Excel	None	1991-2001	Spot market	AAFC	Various	~500,000
		2	Settler	None	1994-2001	Spot market, input packages commodity futures	Svalof. Weibull & Pioneer Hi-Bred	Value Added Seeds	<200,000
GM input trait		3	Innovator; Independence; 3850 and 3880	Liberty-link gene	1995-1996	Management contract with Pools; grower contracts; CCC activity	AAFC and PGS for AgrEvo	Sask, Alberta and Manitoba Pools	215,000
Non-GM output traits	Variety standards and contract registration rules in Seeds Act and trademark for canola	4	DMS-100	Specialty oil	1997-2001	Contracts with Pioneer Grain and Canbra plus grower contracts	Dow Agro Sciences	Pioneer Grain (JRI)	<150,000
GM output trait		5	LA 161; LA269	Laurate gene	1997-1999	Grower contracts	Calgene/ Sask. Wheat Pool	Sask. Wheat Pool	<30,000

Table 8.3: Illustrative examples of types of supply chains in the Canadian canola industry, 1990-2001, *continued*

	Growers	Grower premium	Transport	Crushers	Marketing arrangement
Non-GM input traits	~5,000	None	Growers truck to crusher or elevator; grain co manages rail	4 companies, 7 locations in Western Canada	Seed, oil, meal shipped to any of 35 markets by various agents
	<2,000	Financing	Growers truck to crusher or elevator; grain co manages rail	4 companies, 7 locations in Western Canada	Value Added Seeds arranged sale of seed either domestically or offshore
GM input trait	<2,700	Est. C$2/acre savings on inputs	Commercial trucking to crusher arranged by Pools	Canbra at Lethbridge, CanAmera at Altona and Harrowby	XCAN directed oil and meal to North American market
Non-GM output traits	<1,000	C$40/t + C$2.50/t storage costs	Commercial trucking to crusher arranged by Pioneer, Canbra and growers	Canbra Foods at Lethbridge	Pioneer Grain (JRI) arranged sale of seed to Japan
GM output trait	<300	C$35/t	Commercial trucking to crusher arranged by Saskatchewan Wheat Pool	CanAmera at Altona	Calgene arranged sale of oil to US; meal was used as green manure or destroyed

Note: t = metric tonne; CCC = Canola Council of Canada; AAFC = Agriculture and Agri-Food Canada; PGS = Plant Genetics Systems; JRI = James Richardson International; and Pools = the combined actions of the Alberta Wheat Pool, the Saskatchewan Wheat Pool and the Manitoba Pool Elevator.

Sources: 1: Author's estimates; 2: Allen, 1999; 3: Kennedy, K., 1999; 4: Kennedy, B., 1999; and 5: Saskatchewan Wheat Pool, 1997.

Supply Chains for Input-Trait Crops

The three models of supply chains for crops with input traits operating in Canada during the 1990s are all voluntary. The first type is based on spot markets throughout the supply chain. This is a non-proprietary system, which for the most part involves public or non-proprietary varieties (e.g. AC-Excel), uses spot markets to effect the transfer of inputs and product and conforms with the public grading system established through the Seeds Act and the Canadian Grain Commission.

The second type of supply chain that emerged during the 1990s involved a wide range of proprietary producer contract systems. In these cases, such as for Settler, a grain merchant (e.g. Value Added Seeds) acquires access to a new variety that has the potential to gain market share and makes that variety available to growers only under a production input and delivery contract. In short, the grower buys a package of inputs (seed, fertiliser and herbicide) from the grain merchant, which finances the transaction until the grower delivers the resulting harvest to the merchant. The objective of this type of arrangement is to lock-in input sales and output volumes. In the late 1990s, these types of contracts were worth as much as C$70/acre. These contracts tended to be open-ended pricing arrangements, with provisions for growers to lock in delivery prices based on futures prices. Unlike many of the other contracts in the industry, the grower obligations are usually limited to the specific product contracted and do not restrict in any way the production of other varieties of canola on the growers' farm. In effect, these contracts exist only where there are both corporate benefits and farmer returns.

The third system evolved in 1995 with the introduction of two HT varieties of canola. Under Canadian law new GM varieties that meet Canadian health, safety, environmental, feed and seed regulations are approved for unconfined commercial release regardless of any potential market difficulties. In 1995 Canadian approval came before the key export markets had approved the seed for importation. Earlier research (Smyth and Phillips, 2001) documented the IPPM systems that evolved to handle the market risks of these new varieties (Table 8.4). Between 1995 and 1999, Monsanto and AgrEvo (now owned by Bayer Cropscience) were involved in five systems that have differentiated ten GM varieties on approximately 385,000 acres.

Each of the supply chains began with a specific variety which included a proprietary herbicide tolerant gene which was backcrossed or inserted into a plant by either a contract breeder or by a partner company (e.g. Agriculture and Agri-Food Canada, Plant Genetics System, University of Alberta, Alberta Wheat Pool, Limagrain, Pioneer Hi-Bred or Zeneca/Advanta). Once this variety was registered, Monsanto or AgrEvo contracted with one of the grain merchants to manage the development and management of an IPPM system. That company then multiplied the seed, undertook production contracts with specific farmers, arranged delivery from farms to a processor with contract truckers, and arranged for a custom crush and diversion of the resulting oil and meal into the North American market. As the objective of the IPPM system was to differentiate the HT canola from traditional canola marketing channels, none of the GM canola could touch any part of the export handling system, including elevators, rail cars or port terminals. The differentiated GM production was delivered to Canadian oilseed crushing plants that had markets for the oil and meal in Canada and the US, where regulatory approval had been granted (Saskatchewan Wheat Pool, 1997). In each case, the grain merchant acted as the operating agent for the system, managing the supply chain from seed multiplication to processing.

The participating companies all agreed that the herbicide resistant technology brought real value to producers and all agreed that there was a need to bring this technology to the marketplace. Both AgrEvo and Monsanto acknowledged, however, that if these varieties were co-mingled in the export system, then Canadian canola would be shut out of export markets.

In response, private Canadian firms agreed to release materials only if they were approved in the 'key canola markets', defined as Canada, Japan, US and Mexico by the Expert Committee for Canola of the Canadian Pest Management Review Agency.

Cost estimates were found for two of the five IPPM systems (Table 8.5). While there is room for debate about the actual numbers, the two estimates suggest that transaction costs for the IPPM system were very high. There are five main areas where additional costs were incurred: by the producer on-farm; during transportation; by handlers; by the processor; and through opportunity costs.

The added cost for producers was due to separate storage requirements of the IPPM system. Farmers were required to store these varieties in separate bins, which at times left some of the bin capacity unused. The cost of inefficient use of on-farm storage was estimated to be C$1/metric tonne. Transportation costs were estimated to be higher due to dead freight and freight inefficiencies. Dead freight costs relate to the volumes on-farm that had to be delivered as partial loads. Freight inefficiencies are defined as those costs of trucking that exceeded average costs of delivering to the nearest local elevator. A producer would normally deliver canola to their nearest local elevator, but the canola had to be trucked directly to the crushing facility because elevators were excluded in this system. The inefficiency in transporting the transgenic canola was more substantial as only a subset of crushers were used. AgrEvo used the CanAmera sites located in Manitoba and Alberta while Monsanto used a wider array of crushers in Saskatchewan. Nevertheless, each system used a specific set of crushers. With an average of 40%-45% of canola production in Saskatchewan, producers often faced lengthy trucking distances. These costs were shared among the producers, seed companies and the grain elevator companies. The freight inefficiency cost of transporting the HT canola to processors resulted in costs of C$7-10/t. Dead freight costs were estimated to be C$2-3/t.

The canola processors faced cost increases due to the IPPM system. Manitoba Pool Elevator (1996, p. 2) documents note that "the domestic crusher was obliged to differentiate raw transgenic canola seed, transgenic canola oil and transgenic canola meal from traditional stocks under the [identity preserved production] IPP system developed for transgenic canola introduction." This required the processors to physically clean the production equipment prior to crushing the HT canola (usually during a seasonal shutdown), as well as after the HT run was finished. This was done to ensure that the transgenic canola oil did not co-mingle with oils destined for export markets. The processors involved identified their incremental cost to be in the range of C$3-5/t.

Table 8.4: Key elements in the canola-based IPPM systems, 1995-99

Links in the supply chain	AgrEvo LibertyLink™ varieties			Monsanto Roundup Ready^R varieties	
Species	B.napus	B.napus	B.napus	B.rapa	B.rapa
Variety names (year approved)	Innovator (95)	Quantum (95) and Quest (96)	LG3295 (96)	41P50 and 41P51 (96)	Hysyn 101 RR (97)
Seed developer	AAFC and Plant Genetics Systems with AgrEvo	University of Alberta and Alberta Wheat Pool for 3 Pools	Limagrain	Pioneer Hi-Bred	Zeneca/ Advanta
Grain merchant	Pools	Pools	Cargill	United	Cargill
Farmers*					
1995	310	480	0	0	0
1996	2,375	1,700	incl	0	0
1997	0	0	0	minimal	incl
1998	0	0	0	625	incl
1999	0	0	0	800	incl
Trucking arranged by:	Pools	Pools	Cargill	United Grain Growers	Cargill
Crushers	Canbra at Lethbridge, CanAmera at Altona and Harrowby	CanAmera at Nipawin and Lloydminster	Cargill at Clavet	Archer Daniels Midland (ADM) at Lloydminster	Canbra at Leth-bridge

Pools = Alberta Wheat Pool, Saskatchewan Wheat Pool and Manitoba Pool Elevator.

Sources: Evans 1999, Button 1999, * calculated by authors.

Many non-recoverable costs occurred in administration. The Saskatchewan Wheat Pool, for example, actively managed all of its producer contracts to ensure compliance with the terms and that co-mingling was avoided. In an effort to ensure the purity of the IPPM system, seed agents from AgrEvo, Monsanto and the Saskatchewan Wheat Pool (Pool) mapped all the fields in which HT canola was grown. Once harvest commenced, AgrEvo, Monsanto and the Pool co-ordinated to ensure that an agent from one of the companies would be on-farm during the harvest to inspect the harvested supplies, apply grain confetti and to seal the bin. The paper confetti (similar to that used at many weddings) had the company logo on one side and a unique grower identification number on the other. This placed a very real constraint on producers while at the same time providing the Pool marketing department with accurate information on where and how much HT canola was available. At this point, the Pool worked with the processors to arrange to have the HT canola trucked to be crushed at a designated oil crushing plant that was just about to have a scheduled shut down and cleaning. Once a crush date was determined, the Pool contracted with commercial truckers to pick up the HT canola from farmers and deliver the HT canola to the designated processors. When the canola was to be trucked, an agent from the Pool was again on-farm to inspect to ensure that none of the bins had been opened or tampered with. This process was very difficult due to the simple logistics of trucking grain in the winter in Western Canada. Snowstorms, impassable roads and bad driving conditions all complicated the co-ordination of the trucking process. These requirements were all labour intensive and were estimated to cost C\$4-5/t.

Table 8.5: Identified costs of 1996 IPPM system for HT canola in Canada (C\$)

Cost Category	AgrEvo & Manitoba Pool Elevators ($/t)	Saskatchewan Wheat Pool ($/t)
On-farm costs	$1	$1
Freight Inefficiency	$5-6	$7-10
Dead Freight	$1.50-2	$2-3
Processor	$3-4	$3-5
Administration	$4	$5
Opportunity cost	$20	$10
Collective subsidy		$5-7
Total IPP Cost	**$34-37**	**$33-41**
t = metric tonne		

Source: Manitoba Pool Elevators 1996; Saskatchewan Wheat Pool, 1997.

The grain merchants also identified an opportunity cost of crops in IPPM systems. In effect, differentiating the seed and being constrained on when and where to bring it to the market, and being forced to move it according to some predetermined plan, severely limited the marketers in trying to lock-in high prices in what is traditionally a volatile market. Saskatchewan Wheat Pool (1997, p. 1) reported that "from a general market perspective, an IPP program like this does not allow for access to all attractive alternative markets. This is a potential cost due to possible increased margin potentials, which cannot be achieved. The potential unrealised profit opportunity could well be in excess of $10/mt." Some of that opportunity cost could have arisen because North American buyers recognised that they had some market power in the circumstances and exploited it. Furthermore, the grain merchants estimated other unallocated expenses cost all parts of the supply chain an estimated C$5-7/t. The Saskatchewan Wheat Pool (1997, p. 3) concluded that "in order to develop and promote this technology, the producer of the technology, AgrEvo [and Monsanto], the producers of the seed, the [Saskatchewan Wheat Pool, Alberta Wheat Pool and Manitoba Pool Elevators], and the beneficiary of the technology, the producer, all contributed to the subsidisation of the [IPPM] program."

These cost increases, however, must be balanced against any expected or realised gains. Early figures suggested that farmers gained upwards of C$10/acre or C$5/tonne benefit from the new technologies (Mayer, 1997) and the Canola Council of Canada (2001) estimates that the net benefit to farmers between 1997 and 2000 was C$464 million, which for most farmers would have more than compensated for their added producer costs. The grain merchants may have gained margins on new volumes since then, which probably have compensated for their incremental system costs: by 2000 an estimated 88% of canola acreage used HT varieties. Furthermore, in most cases the canola was crushed through subsidiary crushers, increasing their volumes and offsetting some of the incremental costs. Finally, although the research/seed companies may have lost some money due to the costs of the IPPM systems, they gained significantly in terms of market adoption. It is very likely that if Monsanto and AgrEvo had introduced the seeds without IPPM systems, farmers would have shunned the new seeds. Hence, the only real choice to encourage early and aggressive adoption was to pay for IPPM systems. In this case, one could argue that the two companies accelerated adoption by at least one year, which was estimated to have a net present value in 1995 of more than C$100 million (Table 8.6). Clearly, the IPPM system for HT canola directed to Japan was a win-win strategy.

The results are less clear for IPPM systems for either the EU or for *B. rapa*. Given the uncertainty about the EU, the canola industry made a deliberate decision to abandon the market in 1996. The EU market was not a strong export market for Canadian canola—the EU is usually self-sufficient in terms of canola production and actually exports canola when price premiums are available. The canola industry's view was that future canola exports to the EU had limited potential so, once Japan approved GM products, the IPPM system was discontinued, thereby removing all possibility of supplying EU canola markets.

Table 8.6: Net present value to innovators of early adoption of HT canola varieties in Canada

	% acres in GM canola		Revenue impacts of one year delay in introduction, assuming C$15/acre benefit (C$M)			
	Actual	Delayed	Actual	Delayed	Absolute difference	Net present value in 1995 of difference
1995	1%		2	0	-2	-2
1996	4%	1%	5	1	-4	-4
1997	33%	4%	60	7	-52	-43
1998	44%	33%	89	67	-22	-17
1999	69%	44%	143.9	91.7	-52	-36
Total			300	167	-133	-101

Sources: GM adoption rates from National Research Council, 1998; Canola Council of Canada for acreage; discount rate is 10%; Authors' calculations

Monsanto's IPPM system that has been used since 1997 for four *B. rapa* varieties has posed more difficulties. Although the logic is the same as for the *B. napus* varieties, the economics is different. *B. rapa* has taken longer to get approval because of different environmental impacts (the seed can stay in soil for up to 15 years and still germinate). As a result of the longer approval time, and much smaller market (*B. rapa* is estimated to account for less than 10% of canola acreage in Western Canada in 1999) Monsanto decided,to terminate its work on *B. rapa*.

Supply Chains for Output-Trait Crops

The two main types of supply chains for crops with output traits operating in Canada during the 1990s were required by law because of the potential health and safety risks for the novel traits involved. A few IPPM systems for proprietary non-novel trait varieties have operated (i.e. not mandated by law) but no details are available.

The fourth type of supply chain is designed to handle both commercial interests and public health and safety concerns about the novel attributes in the varieties, in order to allow the production of both food grade canola and industrial rapeseed at the same time. The fundamental driver for this system is food safety. Some products, such as high erucic acid varieties and Dow AgroSciences' DSM-100, which has low linolenic and high oleic acids, would contaminate the food chain if co-mingled and are therefore mandated by law (via Contract Registration under the Canadian Seeds Act) to be produced under segregation rules. Dow, for instance, developed, through traditional breeding processes, a variety of canola that has a novel oil output trait designed to meet processor and consumer demand for an edible oil product that is low in saturated fats. They developed the seed themselves and introduced it to the marketplace in 1997 through marketing arrangements with Pioneer Grain. Dow contracted with Pioneer to multiply the seed and to introduce the seed into the market through production contracts, which for DSM-100 specified the inputs to be used, compulsory delivery, a C$40/tonne premium to producers, a C$2.50/tonne producer storage subsidy for late season deliveries (paid by the Japanese importer) and restrictions on other canola crops on the land. In addition, the grain merchant assembled the resulting crop and arranged export of the product. This system was explicitly designed to ensure delivery of product, which met or exceeded the end customer specifications. Although not the primary objective, the company admits that the IPPM system provides a vehicle to protect their intellectual property. During the first three years of operation, the company asserts it has been able to both lower costs and improve efficiencies on an annual basis. Dow believes that the skills and techniques they are developing from their experiences have provided value and help to create a competitive advantage.

The fifth type of supply chain, which involves a canola variety that possesses both genetic modifications and novel industrial output traits, is similar to the fourth type. Calgene's relationship with the Saskatchewan Wheat Pool to manage the production, delivery and crushing of Laurical™ canola varieties illustrates the challenges of this type of system. As with DSM-100, laurate canola is mandated by law to be produced under segregation. Being a 'novel trait' variety, the seed required additional testing for environmental, food and feed safety impact compared with DSM-100, with the result that the development costs were raised. Meanwhile, laurate is a relatively low-value oil and Calgene and the Saskatchewan Wheat Pool discovered that the sum of the C$35/tonne grower premium (paid to compensate for lower yield and tight restrictions on other plantings of canola) and the incremental IPPM costs more than offset any premium for the oil, with the result that the variety was not produced in 1999.

Supply Chain Governance Mechanisms

Quality in the global grains and oilseeds industry historically has been managed and protected through a wide variety of open access horizontal mechanisms. Increasingly, however, these systems are being supplemented and at times supplanted by proprietary vertical relationships. So far the new systems have facilitated new product development, but at times in an unsustainable way.

The traditional governance system is based on an extensive horizontally-based public/private regulatory system (Altman and Phillips, 2001). In Canada, for instance, the Seeds Act is the first point of quality assurance, as new varieties must on average at least equal the quality of previous varieties. This is administered by the Western Canadian Canola Rapeseed Recommendation Committee (WCCRRC), a committee of more than 30 public and private breeders which evaluate new varieties against the 'check' varieties and recommend varieties for release. This standard has been backstopped by the Canola Council of Canada trademark on canola, which specifies that products must have at most 2% erucic acid and 30 micromoles of glucosinolates per 100 grams of dried meal. Furthermore, the new variety approval system periodically raises the bar for new varieties by choosing a new 'check' variety as the base for standards, which sets the base for oil and meal properties, yields and disease resistance. Once the varieties are approved, the Canadian Seed Trade Association manages the seed multiplication system, specifying the tolerances for substandard materials, and the retail seed business, by overseeing the sale of seeds by registered name. The Canola Council of Canada, the provincial growers associations and various provincial government agencies furthermore provided extensive agronomic extension advice to growers during the season. After the harvest, the Canadian Grain Commission takes over quality assurance for much of the product, setting and enforcing grades and standards for the trade. The Canola Council of Canada, a not-for-profit industry association involving growers, grain merchants, crushers and exporters, ultimately oversaw the evolution of the entire system by licensing and defending its use of the canola trademark. Within this context, spot markets have relatively efficiently managed the commercialisation of a large number of new varieties over the years (Kennett *et al.*, 1998).

As the spot market has been supplanted by new supply chains with closer relationships along the chain, quality is increasingly being managed by and for private interests. The growers' contract, which provides the base for all of these new systems, for instance, specifies a variety of obligations and quality standards that manage the value of the new product within the supply chain. This has included agronomic advice tailored to specific proprietary varieties. Meanwhile, vertical relationships extend back from the grain merchant to the seed and research companies and, in some instances, forward from the crusher into the processed foods sector. This trend has been supported with recent changes in the Canadian regulatory system, for example, which allows for accelerated contract registration for new proprietary novel-trait varieties without going through full review by the WCCRRC.

As new proprietary supply chains have proliferated, a number of difficulties have surfaced. In interviews, company officials note that there is a critical need to have the ability to rapidly analyse grower deliveries at the consolidation point to ensure the product meets specifications and is not co-mingled with other materials. Given the traits involved, the companies are unable to adequately assure quality. Furthermore, there needs to be more capability to differentiate various products into more manageable and cost effective units; most companies agree that many of the new products being considered cannot be justified if the differentiation system continues to cost up to C\$40/tonne. The majority of the new structures being developed (both elevator buildings and handling systems) are not cost efficient in handling small-lot, non-commodity movements. Farmer training also remains a challenge. Teaching growers to have a 'quality mindset' versus a 'quantity mindset' is a key challenge in agriculture in most developed countries. Although the information required to assist farmers with growing specific varieties is often only available from the research companies, the focus on quality requires some horizontal effort. Finally, it is interesting to note that, thus far, industry-based quality assurance has been very limited. For example, apart from the efforts of a few commodity groups (e.g. the American Soybean Association and Canola Council of Canada), there has been little focused effort on developing common quality assurance systems with credible third-party audits.

Conclusions

Market transactions for goods with experience and credence attributes require a high degree of trust, which requires both effective public and private regulatory mechanisms. The canola industry's experience with genetically modified herbicide tolerant varieties illustrates that where there is a public base for managing credence and experiential issues, the industry can effectively handle many of the market considerations through identity-preserved production and marketing systems. Provided the expected returns exceed the costs, private initiative will work. All industry participants assert that this will depend on tolerance levels for shipments (Kennedy, B., 1999). Regardless of whether an IPPM system is established to capture value for a GM trait, special crop trait or traditional variety, it can not deliver a 100% guarantee of purity. Realistic tolerance levels will need to be implemented prior to the increased use of IPPM systems.

Provided economically tractable tolerance levels can be established, IPPM systems may become a permanent method of capturing attribute value from agrifood product markets. Kennett *et al.* (1998) observed that grading standards can reduce the need for vertical integration, which is probably true for search and experience goods. Credence goods, however, impose requirements that a grading system cannot handle. Industry participants in those IPPM systems studied observed that the design of every IPPM system will vary depending on the genetics and marketing of the variety involved. Grading, which homogenises products, would not satisfy the commercial needs of the industry.

If IPPM continues to be required for regulatory and market reasons, it will need to become more efficient. IPPM systems technically work for smaller scale production but it is unclear whether they would work for larger scale operations. While some stakeholders believe that if an IPPM system were spread over a much larger production, efficiencies would be possible. Others believe that there are too many supply constraints (e.g. trucking, storage) for it to work. While there is room to debate the cost estimates provided by industry, IPPM systems could continue to cost in the C$30-40/tonne range if quantities remain small and systems purpose built. If, as was the case in this example, the technology does not impart any perceived consumer benefit, all of those incremental costs will need to be borne in the supply chain, which will probably slow investment in input traits. The focus would probably shift towards seeking output traits that have added value to consumers that can help to compensate for the higher costs.

So far all of the IPPM systems developed have been custom built to meet the specifications of the technology owner and the market. The limited horizontal co-ordination between the systems has come through the research companies (e.g. Monsanto and AgrEvo) working with their agents (the grain companies) and through the commodity groups efforts in export markets. For the most part the grain companies have viewed IPPM systems as valuable proprietary services. Ultimately, however, these systems are designed to earn trust, which is a cumulative process. Past successful actions can contribute to achieving a higher level of trust but failures in one part of the market can spill-over to other market segments. If IPPM systems are here to stay, then it may not be enough to rely on independent systems.

There would appear to be two ways in which IPPM systems could be made more efficient. Ultimately, the goal should be to manage risk. One of the pan-industry participants—such as commodity groups—could become the custom developer or 'integrator' for the system, providing purpose-built but quality-assured systems to meet market needs. Alternatively, the industry could adopt an external quality assurance system such as ISO or HACCP systems to standardise the process of developing IPPM systems. This would differ from a traditional grading system in that the quality assurance system would assure integrity of process and not the standard itself, which would be a negotiated or contracted feature determined in the marketplace. This would leave the operation of the system in the hands of the firms with equity at stake, but at the same time help to build cumulative trust in the system. The difficulty is that neither entity has any equity at stake in the transactions, which might reduce their credibility in the eyes of producers and customers.

As the demands for differentiation change, so too will the design specifications of IPPM systems. From this research, it is clear that differentiation systems require leaders or 'integrators'. The supply chain can be organised and regulated by an integrator at the start of the supply chain or at the demand end (as appears to be happening in the UK). Furthermore, this research suggests that regardless of where the initiative starts, it is extremely important that that integrator has a financial stake in the outcome. Hence, demand side efforts will require a retailer or manufacturer who can see some benefit, such as a price premium or an increased market share from their activities. The costs are simply too large to do this as a *pro bono* effort.

Ultimately, new supply chains have the potential to either control and manage socio-economic liabilities, or to compound and exacerbate the size and impact of these failures. The challenge is that markets cannot in and of themselves succeed. As discussed in the context of many of the new supply chains evolving in Canada and elsewhere, governments play a critical role in defining the relationships within supply chains or between the supply chain and consumers. Chapter 10 provides a clear illustration of the challenge. As plant-made pharmaceuticals evolve, it will be critical for both governments and industry to be clear about their respective objectives, roles and responsibilities. Lack of clarity is a recipe for failure, which as argued above, would cause significant repercussions in many other markets.

Chapter Nine:

Product Differentiation Strategies

Introduction

Agricultural biotechnology has in a decade dramatically impacted the supply and demand of agricultural food products. While biotechnology has offered few new food products to the marketplace, it has revolutionised the method of producing and delivering conventional food products. As discussed in Chapter 8, an increasing number of cereals and oilseeds are being differentiated to ensure that their value or uniqueness is captured and maintained throughout the supply chain. As product demands abound, the potential for failures and corresponding liability rises.

At present, over twenty-eight countries plus the EU have either developed or publicly declared their intent to introduce mandatory labelling legislation for genetically modified products (Phillips and McNeill, 2002). If exporters in countries producing GM crops wish to retain those export markets requiring labelling as clients, then systems of product differentiation will have to be established to ensure the continuity of exports to these concerned markets. The potential for the rapid development of liabilities exists if exports within countries producing GM crops attempt to export products to non-GM countries without properly structured product differentiation systems. Importers within these non-GM countries may seek compensation using liability if they process the raw product into processed foods that are tested and found to contain GM ingredients.

One implication of the evidence reviewed in Chapter 8 is that formal governance structures for these differentiation systems are frequently lacking, creating significant risk of failures. Product differentiation systems can be imposed at the time of variety registration if the novel trait is deemed to harm food safety. As noted in Chapter 8, though, more often than not differentiation systems are not required by government edict. Rather, they are undertaken to realise private objectives. However, private firms must take great care to ensure that the product differentiation systems they choose will achieve their objectives.

If, the global food industry does not adopt product differentiation systems, two alternatives are possible. First, the global food market could continue to divide into distinct markets with only limited interaction between increasingly specific segments. Some markets, such as the EU, and some food processors, have, for example, decided to forgo GM technology for now, and are devoting increasing effort to securing adequate volumes of GM-free foodstuffs to satisfy their customers. Consumers in those markets, such as in the UK, for the most part, do not have any opportunity to consume GM foods. They simply are not available, even though a recent poll of British consumers found 40% were indifferent to consuming GM food (MORI, 2002). Other markets, such as in North America, have rapidly adopted the technology and for the most part do not offer a choice of GM-free food to their consumers. Trade between these two blocks has slowed dramatically in all product markets where GM varieties are being used. As a result, GM producers and exporters face the risk of losing markets where GM traits have not been commercialised and do not wish GM foods in any product lines. For example, US maize exports to the EU dropped from US$574 million in 1995 to US$175 million in 1999 and soybean exports dropped to US$1.1 billion from US$2.1 billion while Canada's canola exports to the EU dropped over the same period to C$8 million from C$204 million (Industry Canada, 2000). Meanwhile, food processors in places like the EU have diverted their purchases to markets where GM varieties are not produced, such as northern Brazil for soybeans and Australia for canola.

A second alternative would be for companies to shelve their new technology. So far a number of technologies have been withdrawn in the US and Canada. In both the US and Canada, for example, GM seed potatoes were withdrawn from the market in 2001 in response to food processor concerns while the developer of GM flax in Canada announced it had deregistered the variety because Europe, which imports approximately 65% - 70% of Canadian production, has not approved the variety for import.

Neither of these alternatives is desirable. If exporters, for example, adopt the technology and lose key premium markets, much or all of the benefits of the new technologies will be offset by market losses, with producers facing the greatest losses. Similarly, if exporters forgo productivity or quality enhancing opportunities, its producers will face even stiffer competition in the residual commodity markets (Phillips and Khachatourians, 2001). Lehnert *et al.* (2000) succinctly make the point about successful product introduction:

> Failures occurring during the establishment of a new product or a new
> process cause high correction costs. They may often lead to losses of market
> share and damage to the image of the supply chain. It is therefore reasonable
> to pay attention to potential failures in the early stages of establishment and
> process planning. (p. 409)

This lynch-pin concept defines a new approach required for the introduction of additional GM crops: do no harm! Caution, diligence and concern for others must be the leading motto for all participants in differentiation systems. In the past, the focus has been on getting new products into the market, while adjustments to the supply chain were made as one went along. Clearly, this strategy is risky. Industry needs to identify and learn from the difficulties, successes and failures that occurred when introducing GM canola, maize, cotton and soybeans to ensure the successful introduction of other new GM crops.

If diligence is not taken in the delivery of products destined for the marketplace, co-mingling of product ingredients may occur and result in improper labelling. Both of these actions establish the precedence for the development of liabilities. Efficient use of product differentiation systems can contribute greatly towards the reduction of potential liabilities.

Definitions of Product Differentiation

The definition of product differentiation can have several nuances, depending on the justification for the differentiation. Frequently, the terms identity preserved production and marketing, segregation and traceability are used interchangeably in the biotechnology and supply chain literature. This is creating misconceptions about the distinct role that each of these product differentiation systems has in the supply of food products and alternatively creates potential liabilities. The purpose of this section is to identify definitions that exist in the literature to date and to suggest definitions where the literature is absent.

Identity Preserved Production and Marketing

The first product differentiation system, identity-preserved production and marketing, has evolved over time in the grain and oilseed industry. Purchasers of raw products became more demanding about the quality and purity of the product they were purchasing so the grain handling system gradually developed distinct channels to market the differing grades of product. All grains and oilseeds are purchased under a grading system in today's marketplace and this grading system has premiums that rise as one moves from low to high grades. The relationship of premiums to differing grades for private market incentives is the definition of an IPPM system.

Identity-preserved production and marketing systems are initiated by private firms in the grain and oilseed industry to extract premiums from a marketplace that has expressed a willingness to pay for an identifiable and marketable product trait or feature. An IPPM system is a 'closed loop' channel that facilitates the production and delivery of an assured quality by allowing identification of a commodity from the germplasm or breeding stock to the processed product on a retail shelf (Buckwell *et al.*, 1999; Lin, 2002). These IPPM systems are predominantly voluntary, private, firm-based initiatives that range between systems that are loosely structured (e.g. malting barley) with high tolerance levels and those with rigid structures (e.g. non-GM produce for European markets) with minimal tolerance levels. Firms operating in the minimal tolerance field achieve this by developing and adhering to strict protocols that specify production standards, provide for sampling and ensure appropriate documentation to audit the flow of product.

A survey of the literature on IPPM shows that while there is growing discussion about IPPM systems, there are very few working definitions. Lin (2002, p. 263) suggests that an identity preservation system "... is a more stringent (and expensive) handling process and requires that strict separation, typically involving containerised shipping, is maintained at all times. Identity preservation lessens the need for additional testing as control of the commodity changes hands, and it lowers liability and risk of biotech and non-biotech commingling for growers and handlers." This definition conflicts with the definition offered here as Lin sees IPPM as having a limited role in the movement of grains and oilseeds due to extremely low tolerance levels. Lin's definition of IPPM and segregation still deal with the same system, one that is initiated voluntarily by private firms in an attempt to capture premiums. It is shown below how IPPM systems differ from segregation systems.

The remainder of the literature on IPPM systems relates to theoretical and operational uses of IPPM systems. Bullock *et al.* (2000) and Bullock and Desquilbet (2001) discuss differentiation between GM and non-GM products and Herrman *et al.* (1999) examine the feasibility of wheat segregation. Bender *et al.* (1999), Bender and Hill (2000) and Good *et al.* (2000) have released a series of papers on handling speciality maize and soybean crops, with costs being the focus, not the defining of the system used to handle the speciality crop. Additionally, Miranowski *et al.* (1999) offer some perspectives on the economics of IPPM, while Maltsbarger and Kalaitzandonakes (2000) provide a solid theoretical model for examining the cost of identity preservation. Moss *et al.* (2004) use an empirical model in an attempt to identify the costs of identity preservation when differentiating between GM and non-GM markets.

Numerous IPPM systems are operating around the world. Some extend only between the breeders and the wholesale market or processor, while others extend right up to the retailer. Their structure depends on the attribute they are trying to preserve. Some novel oils, such as low linolenic oils that are more stable in fryers, only have value at the processing level while others, such as high oleic oils, have health attributes that can be marketed to consumers. Identity preserved production and marketing systems are important for providing information to consumers about the provenance of a product, as those attributes are not visible or detectable in the product itself.

Organic products are one of the most noticeable IPPM products in today's marketplace. Others include: Cargill's IPPM system for the export to Japan of an Intermountain Canola variety that gives off virtually no odour when used to fry food; General Mills IPPM system for a select variety of white wheat that possesses a special trait for 'flake curling' when processed into breakfast cereal; and DowAgro Sciences export programme for Nexera canola to Japan where it is sold into the speciality gift oil market. This chapter documents the Warburtons wheat IPPM system, one of the larger IPPM systems which has been in operation in Canada since 1995.

Segregation

A second product differentiation system, segregation, has frequently been confused with the grading of different classes of grains and oilseeds in order to receive a higher price for the commodity than if it were allowed to be co-mingled. Segregation systems have a formal structure and in fact can act as regulatory standards. Segregation differs from IPPM in that the focus of the system is not on capturing premiums but rather on ensuring that potentially hazardous crops are prevented from entering supply chains that have products destined for human consumption.

Segregation can be defined as a regulatory tool that is required for varieties that, while approved and commercially released, could enter the supply chain and create the potential for serious health hazards. Segregation systems are usually developed as part of a variety registration process, where government regulators use contract registration to ensure that certain novel varieties do not enter the supply channels of like varieties. The private firm seeking registration of the novel variety has to demonstrate that there is a segregation system developed to ensure the containment of the variety.

Lin (2002, p. 263) defines segregation as the requirement "… that crops be kept separate to avoid commingling during planting, harvesting, loading and unloading, storage and transport." Segregation systems are used when potential food safety concerns exist over the co-mingling of the segregated product and all other like products. In a recent paper, Lin and Johnson (2004) estimate the cost of segregating non-GM maize and soybeans at 12% of the average farm price. In short, IPPM are used to capture premiums and segregation is used to ensure food safety.

There are very few segregation systems presently in operation. In Canada, the best known segregation system is for high erucic acid rapeseed, which has been produced using a segregation system since 1982. This chapter examines the costs and benefits of this segregation system.

Traceability

The third product differentiation system, traceability, is commonly used in the food industry. Retail products found with unacceptable bacteria levels or intolerable levels of pesticide or chemical residues need to be quickly and completely removed from shop shelves. Traceability systems allow for retailers and the supply chain to identify the source of contamination and thereby initiate procedures to remedy the situation.

The key focus of traceability is on greater acceptance and food safety. Recently, the focus for developing traceability systems for new sectors of the marketplace has shifted from food safety towards extracting premiums from the marketplace. But market premiums are unlikely to be large enough to sustain a traceability system, as traceability systems do not in and of themselves define quality, they simply trace it. If market premiums are the driver, then the developers need to use an IPPM system, as they are the only systems properly structured to capture premiums.

The International Organisation for Standardisation (ISO) has defined traceability as the "… ability to trace the history, application or location of an entity by means of recorded identifications …." and the Codex Alimentarius Commission has adopted this as their working definition for all Codex standards (Codex, 2001). The EU (2001) has defined traceability quite clearly in relation to GM products. Directive 2001/18/EC (p. 2) defines traceability as:

> The ability to trace GMOs and products produced from GMOs at all stages of the placing on the market throughout the production and distribution chains facilitating quality control and also the possibility to withdraw products. Importantly, effective traceability provides a 'safety net' should any unforeseen adverse effects be established.

Price *et al.* (2004) provide a solid over view of European traceability as it relates to the American soybean industry.

The economic literature from supply chain management defines traceability as the information system necessary to provide the history of a product or a process from origin to point of final sale (Wilson and Clarke, 1998; Jack *et al.*, 1998; Timon and O'Reilly, 1998). While Dickinson and Bailey (2001) suggest that their results from a laboratory auction market regarding features of meat traceability show there is willingness by consumers to pay premiums for traceability, the key focus has to be on food safety. Prior to adopting traceability systems there has to be a clear indication of specifically what aspects of food safety can be improved by the adoption. Marginal improvements in food safety would be a dubious reason for proceeding—rather there must be a clear and evident improvement in the level of food safety.

Traceability systems have been developed for beef products in many countries around the world. Traceability has been developed in conjunction with a quality assurance system to reassure export markets about the quality of meat products (Spriggs and Isaac, 2001). However, it should be noted that these systems have been met with some resistance at the farm level in some countries, as producers often do not want to allow government regulators onto their farms or provide regulators with any sensitive farm information. In a similar quality assurance effort, the Canadian grain and oilseed industries are conducting a two-year pilot project in 2002 and 2003 to evaluate the costs and benefits of an on-farm HACCP-based traceability system. This chapter examines the implications of this proposed traceability system.

Options for Product Differentiation

Each product differentiation system has features that are unique, while also possessing features that are common to one, if not both, of the other systems. Table 9.1 - 9.4 compares numerous features of product differentiation. These features are classified into those that apply to the complete supply chain and those that apply to the three distinct stages of supply chains. The first stage contains features that are most commonly related to the production stage of the supply chain. Included in this stage are seed development firms, producers and grain handlers. The second stage of the supply chain is the processing stage. This stage includes all firms involved in the supply chain from the point when a raw ingredient is received to the point that a final product is shipped to the retailer. The third stage is the retail stage of the supply chain. This stage includes those firms that provide products to consumers, such as grocery stores and restaurants.

Overall Supply Chain Management

The features in this stage are those that are important to the entire supply chain (Table 9.1). Unlike the features in the following sections, these features span all sectors of the food industry and each participant in the supply chain must ensure that their commitment to these features is at least as strong as the other participants.

Table 9.1: Comparing identity preservation, segregation and traceability in overall management

	IPPM	Segregation	Traceability
Objective	Revenue management	Liability management	Product safety
Status	Voluntary	Mandatory	Voluntary or mandatory
Lead stakeholder	Private company	Regulator	Commodity group, standards organisation or regulator
Regulatory agency involvement	Consumer fraud	Regulatory oversight	Consumer fraud
Information	Asymmetric	Full	Asymmetric
Risk	Moral hazard	None	Moral hazard
Information flow	Two way	Two way	One way
Supply chain focus	Downstream	Downstream	Upstream
Penalties for failure in product market	Consumer fraud charges; lost brand value	Criminal prosecution; mandated product recalls	Consumer fraud charges; exclusion from product category
Testing/ auditing	2nd party/brand owner	1st party/regulator	3rd party/standards organisation

Source: Authors

The objective of an IPPM system is revenue management. Premiums need to be available to attract participants and the efforts of participants will be directed towards receiving a share of the premium. Participation in these systems is inevitably voluntary. The lead stakeholders in IPPM systems are private firms seeking to capture the increased value of special traits. The role of the regulatory body will be to ensure that industry standards are in place to prevent consumer fraud from occurring. The information may be asymmetric as only the product seller can know with certainty what level, if any, of cheating has occurred in the delivery of the product. Moral hazards may be present due to the presence of premiums. Effective IPPM systems that span entire supply chains must have accurate two-way information flows. This means that information about purity and quality of the product flows downstream and that information coming from consumer demand due to identified willingness to pay flows upstream. While the information flow in IPPM systems is two-way, the focus of these systems is downstream. Each participant in the system wants to ensure they are extracting a portion of the value of the special trait whether they are involved with the production, processing or retailing of the product. This means that each participant will focus on the needs of the next participant in the supply chain. Market failure can result in fraud charges for mis-labelling or improper labelling and also create awareness with consumers that certain brand names can not be trusted. Testing and auditing will usually be done by second parties acting on behalf of the brand owner or developer of the special trait.

The objective of a segregation system is to manage any and all liabilities that may arise through the production and processing of a commodity. Participation is not optional—any producer or firm involved with segregated products will have to comply with standards established that have been approved by the regulatory agency. The private firm will have the responsibility of developing the actual system, but the regulatory agency will be the final arbiter on approving the system for field use. Information will be fully disclosed due to the importance of protecting food safety, which will result in the reduction of risks in the system. Segregation systems must have two-way information flow due to compliance with food safety standards. The focus of product delivery within a segregation supply chain will be downstream. Segregated commodities commonly have industrial value, so these products will be supplied to meet the criteria of the processor. The costs of market failure would most definitely see a complete recall of any and all products suspected of being affected. It may also result in criminal prosecution in the most severe instances. Testing and auditing will be vital features of segregation systems and will be conducted by agents of, or acting on behalf of, the regulator. This process will also reinforce the level of trust with foreign export markets.

The objective of traceability systems is to ensure that if unsafe products enter the supply chain, resulting liabilities can be minimised through product recall. Participation in a traceability system can be voluntary or mandatory, depending on where in the supply chain the participant is located. The closer the participant is to the start of the supply chain, the more likely it will be that participation is voluntary. The lead stakeholder may be a commodity group demanding greater clarity in or selection of food products, a standards council that is comprised of industry representatives from all sectors of the supply chain or the regulator to ensure consumer protection. Information may be asymmetric due to the voluntary nature at the start of traceability supply chains. A moral hazard may also exist due to the inability to fully test for some features of traceability. Traceability systems will only have information flows that are one-way as these systems are designed to react quickly to food safety concerns. If a product is discovered to exceed any defined tolerance level at any point in the supply chain, traceability will be used to identify the source of the problem and to locate any and all in-chain and retail products that may be affected. This results in the focus of traceability systems being upstream. Market failures can also result in consumer fraud charges in addition to permanent exclusion from selling into that supply chain. Testing and auditing will be conducted according to the standards developed by third party organisations.

Production Stage Features

The production stage features are those at the front end of the supply chain and involve seed development firms, producers and grain and oilseed handlers (Table 9.2). Historically, this has been the starting point for supply chains, as seed development firms would commercialise a new crop variety and marketing the benefits of the variety would result in producers adopting the variety. This push version of supply chains has had difficulty adapting to the demands of consumers for a pull supply chain.

Table 9.2: Comparing identity preservation, segregation and traceability in production stage features

	IPPM	Segregation	Traceability
Production arrangements	Formal production contracts	Regulation and contracts	Membership in quality standard
Production controls	In-season agronomic rules vary with product	Formal buffer zones; post production land use controls	Process standards adopted and record keeping
Premiums for producers	Short and long term	Short and long term	Short term

Source: Authors

Identity-preserved production and marketing systems are voluntarily developed by private firms to ensure that all stakeholders in the supply chain for a specific product capture a share of the value from special trait varieties. Private firms may use TUAs to protect the intellectual property of the special trait or production contracts that have specific conditions that must be met in order to receive the premium. Grain companies typically organise and manage these contracts. These systems are typically developed for niche market products and are typified by small acreage and low volumes. There is presently some debate as to whether long-run premiums for producers are sustainable, as they may be bid away due to competition among producers.

Segregation is focused on ensuring that the integrity of a special trait is not allowed to adventitiously co-mingle with other products destined for the human food supply. Production contracts would be used by the private firms to ensure that all of the commodity being segregated is collected and that the producer retains no amount of seed. Buffer zones are required for segregation systems as a preventative measure for reducing cross-pollination. Producers may also have restrictions placed on what crop varieties would be allowed to be grown the following year on fields that produced segregated crops. Premiums would be available in both the short and long term to ensure that product supply is maintained.

Traceability is very fragmented at the producer stage. Production arrangements would largely be done through membership in the organisation established to create and manage the industry. Production control would be through industry standards and stringent record keeping. The cost of initially becoming involved in a traceability system results in short term premiums being available to attract producers. Long term benefits are not evident as the premiums evaporate when the desired number of producers become involved.

Processing Stage Features

Processing stage features are those that relate to firms involved in the manufacturing of food products (Table 9.3). Most of these features contain aspects of quality assurance and industry developed standards.

Quality standards in IPPM systems will be enforced by private commitment to industry standards, as the value of the product will be greater given higher purity levels. The enforcement of standards is crucial as products that do not conform to the desired quality level will not be accepted. Tolerance levels will vary from product to product and also will depend on the preferences of the final consumer. Testing and tolerance levels will be important to ensure that the purity and the high quality levels of the product are maintained. Frequently, these tests will be conducted by second parties.

Enforcement of standards will be very important in segregation systems. To ensure that products that could be a hazard to the human food supply chain are prevented from entering that supply chain, functional operating standards must be agreed to by all participants. The enforcement of these standards will need to be rigorous, as it was the lack of standards enforcement by Aventis that resulted in the StarLink™ maize debacle. Quality will be defined in regulations or be created through the implementation of a HACCP system. Tolerance levels for co-mingling will be set by the regulator. Due to the importance of standards, the features of testing and tolerance levels will also be important. Testing will need to be conducted frequently to ensure that the commodity being segregated is being handled properly and that none of the product is entering other supply chains accidentally. This will be done by agents of the regulator.

Table 9.3: Comparing identity preservation, segregation and traceability in processing stage features

	IPPM	Segregation	Traceability
Enforcement	Private	Public	Collective
Quality criteria based on	Product standards	Regulations and or HACCP rules	Processes (e.g. ISO)
Tolerance levels	Variable	Set in law	Performance based
Testing/ auditing	2nd party	1st party	3rd party

Source: Authors

    The processing stage is very important for traceability as this is the stage in the supply chain where traceability begins to be rigorously applied. The lack of high standards and less care or enforcement of the standards results in costly recalls of products, potentially damaging whole product categories. Therefore, the enforcement of standards will be done collectively. Quality will be focused on the production processes to ensure that the highest standards possible are maintained at all times. Tolerance levels exist for food safety reasons as no product can be 100% free of potentially harmful effects, so tolerance levels are established at levels that ensure safe consumption. When tolerance levels are exceeded, the risk of harm to consumers develops and these products must then be recalled from the marketplace. The costs of recall are substantial. Not only does the firm have the cost of gathering and disposing of the product in question, there may also be a loss of trust in that brand name by consumers that will require aggressive marketing campaigns to overcome. Testing and auditing of traceability systems are done by third parties.

Retail Stage Features

The final stage of the supply chain is the retail stage (Table 9.4). The features in this category apply to those firms that are involved with selling food products to consumers. This is the stage of the pull supply chain that is now seen as driving many modern supply chains.

    Identity preserved production and marketing systems may play a large role in the introduction of new GM food products. New GM products often are introduced without complete international market acceptance and IPPM systems can be used to ensure continued access to all markets. An IPPM system will be able to provide information to the consumer about the uniqueness of the branded product that is being identity preserved. For an IPPM system to function properly, and ensure that all stakeholders remain committed to the process, final market price premiums must be available. If this premium is not available for the retailer, an incentive is created for the retailer to no longer carry the product. Products of IPPM systems will need to be labelled to justify the final market premium. If the consumer has no option to identify the value of the product, the consumer will not pay a premium to purchase the product.

Segregation systems will also be used to ensure that market access is continually guaranteed. A coordinated education and marketing effort by the regulatory agency and the private firms involved can be effective in foreign markets in creating trust that production of potentially hazardous products can exist and not jeopardise export market streams. Most segregated products are incorporated into the production process at the processing stage. Thus, there are no final market premiums available, nor is labelling of the product a concern.

Table 9.4: Comparing identity preservation, segregation and traceability in retail stage features

	IPPM	Segregation	Traceability
Provides access to	Branded product markets	Markets	Product categories
Information provided to	Consumer	Regulator	Regulator, retailer or processor
Final market price premiums	Yes	None	None
Labelling	Private brands	None	Quality standard

Source: Authors

Traceability is crucial for providing access to new categories of products. Many markets have demanded documentation regarding product composition prior to allowing market access. Consumer information is fundamentally important for traceability systems as they are designed to increase information regarding food safety to consumers. Information is also provided back up the supply chain to regulators and processors. Final market premiums are not available for traceability systems. Labelling is important to traceability to ensure high quality standards and allow consumers to identify with this feature.

The above attempts to identify features common to different product differentiation systems. These features are classified as to whether and how they pertain to identity preservation, segregation and traceability. As is evident, some features differ depending on the system in which they are applied. It will be important for those involved in product differentiation to examine what features are most commonly related to the product requiring differentiation and if and how those features overlap. This model of comparison will assist with determining which system best relates to the identified needs of the product being differentiated.

Realities of the Marketplace

Using the taxonomy identified in Chapter 1, we can see how the three leading stakeholders involved with product differentiation—regulatory agencies, private firms and civil society groups—relate to each other. Regulators decide to segregate commodities that pose negative health impacts, private firms use IPPM systems to capture value of special traits and civil society groups seek traceability to inform consumers about important health issues.

Section A in Figure 1.1 is typified by commodities that require strict segregation due to concerns about potential health hazards that may arise due to adventitious co-mingling. This process has been used in a variety of countries to mandate the segregation of industrial crops, such as high erucic acid rapeseed. This variety of rapeseed produced in Canada requires segregation because it contains over 50% erucic acid and does not fit with canola, which by definition has to contain less that 2% erucic acid (CGC, 1998). In the US, StarLink™ maize was required to be segregated by the Environmental Protection Agency as this variety had not received regulatory approval for human consumption.

Section B is dominated by private firm initiatives to capture a share of the value associated with a special trait. As pointed out above, the IPPM systems that operate can range from rudimentary systems with high tolerance levels (e.g. malting barley) to rigid systems with minimal tolerance levels (e.g. non-GM shipments to Europe). Firms identify a segment of the marketplace that is willing to pay a premium to consume a product with a specific feature and these firms use IPPM systems to deliver the specified product.

Section C is characterised by products present in the marketplace with traceability systems for food safety concerns. One of the more common uses of traceability systems is in the meat and dairy sectors where product lots with bar codes can be recalled if high bacteria levels are found.

Section D is an area of overlap between the interests of private firms and regulatory agencies. This area is typified by cooperation between the regulatory agency and private firms to ensure that commodities that could be harmful to the bulk or pooled systems can be safely communicated. Systems that develop and operate in this area would be HACCP based as these systems are initiated by the private sector with the close involvement of food safety authorities. Molecular farming, when it arrives, will be associated with this area (see Chapter 10 for a thorough discussion on molecular farming). The regulatory bodies will be very involved in the regulation of molecular farming and will demand segregation while biotechnology firms will want aspects of IPPM systems to ensure that the value is captured.

One example of a Section E case would be mandatory nutrition labelling required by law through regulatory agencies for the benefits of consumer health. Most governments impose labelling rules to require that all food products provide consumers with basic information regarding the content of food products. Some markets are ahead of others in this area. The US FDA, for example, requires mandatory nutritional labelling for all processed food products. The labelling that occurs in this area is not done for immediate health and safety reasons as in Section C, but done for health improvement reasons. By providing nutritional information, it is hoped that consumers will make healthier food purchases.

Section F involves cases where there is an overlap between the interests of private firms and civil advocacy groups. One good example of this would be where firms use enhanced nutritional or product quality to extract premiums from consumers. The systems that operate in this section are largely for small niche market products that have aspects of IPPM, are labelled and adhere to the principles of traceability. Organic, ethically or environmentally produced products and products marketed as GM-free would fit in this category. None of the above products can make any substantiated claims about increased food safety, so they use increased product information as a marketing tool to create price premiums. Quality assurance is an important issue in this area. Private firms use ISO standards as justification for including premiums. Country of origin labelling would also be located in this section as it provides firms with the opportunity to market this information.

Section G is the small area at the centre of the model, where the interests of all three stakeholders overlap. Goods in this section will have aspects of identity preservation, segregation and traceability, such as is increasingly associated with the introduction of GM wheat. It is presently anticipated that the introduction of GM wheat will have greater regulatory requirements than GM canola. The biotechnology industry also hopes that GM wheat will involve price premiums (e.g. Monsanto has publicly pledged to produce GM wheat through an IPPM system). Lately, consumer groups are calling for aspects of traceability for any products resulting from GM wheat.

This attempt to visualise the various methods for product differentiation within the grains and oilseeds market can also be applied to other sectors of agriculture. The role of government within this model may well vary according to the level of trust that consumers have in the product. If consumers have a low level of trust in the private firm or the nature of the product, regulatory agencies may need to become involved by using segregation standards to bolster public confidence. Many in the food industry agree that society is moving towards greater product differentiation, the question that will need to be addressed is 'which method provides the greatest level of benefits, both financial and in terms of information, to the greatest number of participants and consumers?'

Examples of Functioning Product Differentiation Systems

Canadian production practices offer operating examples of each form of product differentiation. The focus of these examples will be to examine the costs and benefits of each system.

Identity-preserved Production and Marketing

The Canadian prairies are known as consistent producers of high-quality milling wheat. The United Kingdom has a long history of importing wheat from Canada and the baking firm Warburtons imports roughly 200,000 tonnes per year. Prior to the 1995 crop season, Warburtons initiated discussions with the Canadian Wheat Board (CWB) and the Canadian Grain Commission (CGC) in an attempt to ensure the purchase of superior qualities of milling wheat (Kennett *et al.*, 1998). An agreement was reached with Manitoba Pool Elevators (now Agricore United) to administer an IPPM programme under the auspices of the CWB. All of the wheat that was purchased through this IPPM system was grown with the use of producer production contracts. In 1995, producers received a premium of C$30/tonne to compensate them for the extra cost and effort of growing contract wheat, and to attract farmers into the programme (Kennett, 1997). The following year the premium dropped to C$20/tonne and has remained at this level (Table 9.5). This premium is designed to compensate producers for the costs of purchasing pedigreed seed and having to store the wheat for longer periods of time than might normally be the case.

Warburtons opened a laboratory and pilot bakery in Brandon, Manitoba in 1996 in an attempt to ensure that the contracts being let met the quality standards demanded by the company (Kennett, 1997). The laboratory tests the quality of the wheat samples that are provided by the producers and also performs bake tests. Investment for testing, whether for quality or purity, was a key component of this IPPM system. In addition to conducting their own testing, Warburtons also pays the standard CGC testing rate of C$4.50/t. The CGC has the responsibility of testing and certifying all wheat exports.

The CWB played a critical role in this IPPM system because of its monopoly over wheat exports. Initially, the CWB was concerned that a system such as this would lead to the 'cherry-picking' of top quality wheat (Kennett, 1997). Once the nature of the programme was determined, the level was deemed to be of such a low volume that this concern dissipated. For its role in the exporting of this wheat, the CWB charges Warburtons an additional premium estimated to be C$2.00-3.00 per tonne (Cooke, 2001) above and beyond the premium provided to producers, which is intended to offset any additional administration or logistical costs caused by the programme (Kennett, 1997). This premium operates as a cost recovery programme. The CWB incurs additional expenses to ensure that this high-quality milling wheat is exported via an IPPM system and, for this, the CWB receives the premium to cover the costs.

The combination of the three premiums that Warburtons pays ranges from C$29.50-C$33.00/tonne. With a 10-year average price for wheat of C$183/tonne, the cost of this IPPM system is 16%-18% above conventional wheat marketing costs. This cost only reflects the costs incurred to have the wheat loaded on ocean-going vessels. Warburtons could experience additional IPPM costs once the wheat reaches the United Kingdom.

The fact that this IPPM system has continuously operated since 1995 would indicate that all the participants are extracting sufficient benefit to continue with the operation of the system. Producers continue to participate as the premium must offset the added costs of certified seed purchase costs and on-farm segregation costs. The CWB's administration costs are covered and they have a growing market that requires no marketing costs to maintain. Agricore United is able to generate additional producer input sales via the contracts. While the management fee is undisclosed (estimated to be C$7.50-C$10/t above the standard handling tariff of C$11/t), one would assume that it must provide enough compensation to Agricore to allow them to continue managing the IPPM system. Beard (2001), however argues that the Western Canadian grain industry is facing such tremendous competitive pressure that some companies are doing things to protect producer market share and customer market share that may not be completely rational from a purely economic perspective. Finally, Warburtons is able to sell a superior quality loaf of bread in the UK at a premium price, thus justifying the producer premium, fee expenses and the quality assurance programme they have implemented. Loaf bread in the UK is typically in the price range of 33p-39p compared to the 86p for Warburtons bread. It would appear that so far the benefits of this IPPM system have been distributed throughout the system, which ensures the continued participation of all stakeholders.

Table 9.5: Costs and benefits of the Warburtons IPPM system (C$)

Participant	Costs	Benefits
Producers	Purchase certified seed annually	$20/t premium
	Specific crop management practices	Grower contract to purchase entire crop
	Additional paperwork	
CWB	Increased administration	Collects premium ($2-3/t est.)
	Reduces quality of remaining wheat	Guaranteed market
Agricore	Writing and monitoring contracts	Certified seed sales
United	Liability for co-mingling	Additional specified input sales
	Under-utilised storage space	Marketing opportunities
	Loss of blending opportunities	Receives fee for IPPM services ($7.50-10/t est.)
Warburtons	Expenses to producers, CWB, Agricore ($29.50-33/t est.)	Higher premiums on sales (45p-50p/loaf or $1-$1.10/loaf)
	Quality control programme including a testing lab in Manitoba	Stable, dependable supplier
		Guaranteed delivery of high-quality milling wheat

Sources: Kennett, 1997; and authors.

Segregation

High erucic acid rapeseed (HEAR) was bred out of conventional rapeseed because of the industrial value of the high erucic acid content. Traditional rapeseed varieties in the 1970s had erucic acid levels that ranged from 30%-40% which were not high enough to be valued for industrial applications (NSERC, 2001). The University of Manitoba developed a rapeseed breeding programme that sought to increase the level of erucic acid to 55%. The first HEAR variety was commercialised in 1982 in conjunction with CanAmera Foods. The erucic acid oil that is produced is valued by the plastics industry for its lubricant qualities.

Producers are required to grow HEAR under contract registration. This requirement was mandated at the time of varietal registration by Agriculture Canada (now the CFIA). Under contract registration, regulators from the CFIA have the right to inspect all HEAR fields to ensure compliance with segregation requirements. This requirement has resulted in some producers not proceeding with production contracts as they do not want to allow government officials access to their farms.

The total annual acreage for HEAR is in the range of 100,000–150,000 acres. Producers receive financial benefits in three distinct forms (Table 9.6). First, they receive a price premium of C$1 per bushel (or C$44 per tonne) above the market price for canola at the time of delivery or contract price lock-in. Second, all producer freight costs (f.o.b. farm) are paid by CanAmera. This would amount to around C$10 per tonne for producers located in Western Canada. Finally, producers are compensated at the rate of C$25 per tonne for dockage. This final compensation results from the very limited weed control options that producers have with this rapeseed variety. Previously, the largest cost for producers has been the size of the buffer zone. The required isolation distance was initially set at 100m, but private research by CanAmera on co-mingling due to cross-pollination with canola has resulted in the distance being lowered to 5m. Producers may still be required to harvest the portion of the crop nearest the HEAR field separately and sell it as animal feed. There is some additional paperwork that is required with the production of HEAR. Producers have to complete post-seeding surveys and map all fields under production. Producers are required to purchase pedigreed seed on an annual basis from the Saskatchewan Wheat Pool (SWP). Furthermore, this variety of rapeseed is very susceptible to most agriculture chemicals and, as a result, most farmers are not able to spray for weed control. This can create weedy fields for the subsequent crop and therefore, raise the cost of subsequent year weed control. Producers are also required to bin all HEAR separately from other crops, which can result in under utilised on-farm storage.

The SWP has been contracted by CanAmera to multiply and sell certified HEAR seed for producers, which provides the grain company with additional seed sales. The incremental costs for the SWP are almost nonexistent, as the only potential cost would be if some canola seed were inadvertently mixed with HEAR seed. If this were to occur, the SWP would have to compensate producers should the level of erucic acid not meet the minimum standard of 47% demanded by CanAmera.

Table 9.6: Costs and benefits of HEAR segregation system

Participant	Costs	Benefits
Producer	5m buffer zone that may have to be harvested separately and sold as feed	C$44/t premium
	Additional paperwork	Freight (f.o.b. farm)
	Purchase certified seed annually	C$25/t compensation for dockage
	Reduced weed control options Inefficient use of storage space	
Saskatchewan Wheat Pool	Liable if the rapeseed seed and canola seed become co-mingled	Certified seed sales
CanAmera	Grain confetti provided to each producer Random audits as required by and conducted by CFIA	Guaranteed industrial market Only processor of high erucic acid rapeseed in Canada
	Crushing facility must be cleaned before and after	
	Producer education	
	Complete policy manual submitted annually to regulatory agency	
	Expenses of C$82-84/tonne	

Source: Slusar, 2002

CanAmera is the only processor of HEAR in Canada, giving them the potential to exercise monopoly pricing powers. By keeping the acreage low, CanAmera has ensured a guaranteed industrial market for the entire annual production of erucic acid. CanAmera oversees the quality of the product by purchasing grain confetti and distributing it to each product contractor. Additionally, any costs associated with the random audits by the CFIA are paid for by CanAmera. The crush of HEAR is co-ordinated to follow a scheduled shut down and cleaning of the processing plant. The HEAR is processed right at the end of a run of conventional canola and then the plant is shut down and cleaned again before it starts to process other oilseeds. The cost of this has been estimated to be C$3-5 per tonne (Smyth and Phillips, 2002). CanAmera has also devoted some resources to assist in educating producers about the mandatory segregation systems for HEAR and the procedures and conditions that will apply if they sign a production contract. Finally, CanAmera is required to annually submit a complete policy manual regarding the structure of the segregation system to the CFIA for review.

The total identifiable costs to CanAmera for this segregation system range between C$82-84 per tonne. Given that the five-year average price for canola in Western Canada was C$346, this results in a cost increase to CanAmera of 24% above the cost of handling conventional canola. The cost of this segregation system may have ranged as high as 30% of the market price in the early to mid 1990s when canola prices were lower. The value that CanAmera receives from the plastics industry for supplying erucic acid must exceed C$85 per tonne to justify continuing the HEAR programme.

Traceability

The demand for traceability systems originated in meat and dairy products. The concern about unsafe bacterial levels of *E. coli* or salmonella resulted in systems to ensure that all products sold to consumers were within the allowable tolerance levels. These systems then moved to fruits and vegetables as concern about chemical and pesticide residues developed. Presently, regulators and industry are examining the merits of transferring traceability systems into the grains and oilseeds sector of agriculture.

Canada is presently in the initial stages of launching a traceability system for grains and oilseeds. The goal of the Canadian On-farm Food Safety (COFFS) programme is to provide assurance to export markets and domestic consumers that the production and marketing of Canadian grains and oilseeds have the highest standards possible. The initiators of this programme are citing the rising level of concerns about quality and food safety coming from foreign export markets as the driver of this programme. The argument being used to justify this programme is that the entire supply chain has to accept responsibility for food safety.

The leading hazards that have been identified are biological, chemical and physical. Biological hazards are those that can cause illness or death due to the presence of microorganisms. The presence of bacteria, yeast, mould, viruses, parasites and mycotoxins in processed food can have serious health implications. Chemical hazards are residues from farm chemicals, cleaning fluids, drugs, lubricants, heavy metals and any naturally occurring toxins. Physical hazards are identified as foreign objects that may cause physical injury to a consumer. Common foreign objects in grains and oilseeds are things like glass, wood, stones, metals, noxious weeds and insect and rodent fragments. The focus of the COFFS programme is to reduce or eliminate the presence of any of these hazards, which are crucial for the marketing of high quality Canadian grains and oilseeds.

Ultimately, this system will be used as a marketing tool by Canadian export firms to extract premiums for IPPM foods. Table 9.7 outlines the costs and benefits of the grains and oilseeds traceability system.

Table 9.7: Costs and benefits of the COFFS traceability system at the producer level

Costs	Benefits
Time devoted to increased record keeping	Premiums for early adopters, but likely to disappear when sufficient numbers have adopted
Infrastructure investment costs to upgrade to meet new standards	Continued market access
Annual certification	
Failure to comply with standards	
Annual audit procedure	
Reduced ability of producer to sell produce if not a member	
Increased potential for liability	

Source: Canadian Grains Council, 2002

The cost to producers will be the main barrier to entry. The cost of documenting and recording every single operation related to planting, spraying and harvesting a crop will be substantial. Many producers will have to make capital investments in terms of new equipment and computers to allow for participation. Annual certification and audits by CFIA regulators will also be a substantial cost. Premiums may be available initially to attract producers to the traceability programme. This was the case in the Scottish Quality Cereals (SQC) programme that is operating in Scotland. To attract producers of malting barley to the SQC programme a premium of 1GBP/tonne was offered in the first year of the programme (Fearne and Garcia, 1999). This premium was discontinued in the second year as the desired number of producers had been attracted to supply high quality malt barley. Producers may ultimately have no choice but to join the programme to ensure that they continue to have market access. If enough foreign markets demand traceability features, all on-farm grain and oilseed sales may have to be done under the standards of the COFFS system.

The failure to comply with predefined HACCP standards may result in producers finding themselves in the situation where their grain or oilseed crop does not meet the traceability standards and, therefore, may have no option to move the commodity. If traceability standards develop to the same level for animal feeds, producers would not be able to sell unacceptable grains or oilseeds into either the food, grain, oilseed or animal feed industries. The only remaining option for the producer would be to burn or otherwise destroy the entire crop. The final cost, increased potential for liability, may in itself be enough to deter participation. Any lawsuit that claims harm from consuming a processed food product would potentially be able to trace the cause of the harm back to the producer of the grain or oilseed. This would result in individual producers being included in litigation suits where substantial financial costs could be awarded for harm and suffering.

Product Differentiation and Liability

The importance of product differentiation is underscored by the recent increase in the influence of consumers in food supply chain practices. The major influence over supply chain development has shifted from those involved in the commercialisation of new crop varieties to those purchasing the final food products. This shift of focus from those at the upstream end of the supply chain to the final actors in the supply chain means that the potential for liability claims resulting from mislabelled food products increases considerably. With consumer influence on the rise, product differentiation becomes central to identifying the proper placing of liability.

Historically, what has happened in the marketplace pertaining to food safety and product recalls is that product quality testing would identify a batch of products that were not in compliance with the production regulations or standards and the product would be removed from grocery shelves to ensure consumers were not purchasing food products that could be harmful. A variety of identification codes have been used to distinguish products and these codes have been very useful when products need to be removed from the marketplace. These inferior quality products have been removed from the supply chain to ensure continual high levels of food safety. What is unique about the identification of transgenic food products is that it is not the fact that transgenic foods are perfectly safe to eat, but that some groups of consumers may be consuming transgenic products without being fully aware of what they are consuming.

Throughout the first decade of consumption of transgenic foods in North America, there has yet to be a single medically documented case of illness attributed to the consumption of transgenic food products. However, North American consumers are becoming increasingly aware of food safety and the importance of dietary requirements for a variety of health related reasons. One result is that the number of people reading labels is beginning to increase. Proper labelling for the presence of transgenic ingredients has the potential to develop into an issue of substantial importance for the retail food industry.

Liabilities can arise from the improper labelling of food products. In some instances, the market liability will simply be a loss of consumer trust in the brand name that produced the mislabelled product. In others, the liability may take the form of frivolous lawsuits which still have to be financed and dealt with by the court system. If a consumer purchases a food product that unknowingly contains transgenic ingredients (even if the consumer does not become ill following consumption of the mislabelled food product), the consumer, particularly US consumers given the tradition of litigation in American society and the generosity of the settlements that can routinely be obtained, may feel compelled to launch a liability lawsuit against the product manufacturer. While it has yet to be scientifically documented that humans can become ill from consuming transgenic food products (knowingly or unknowingly) that will not stop those who perceive themselves as having been negatively affected by consumption of GM foods from attempting to seek financial remuneration for their inconvenience. Aventis faced several consumer lawsuits from the consumption of StarLink™ maize and the cost of settling these lawsuits has been estimated to be US$9 million.

Efficient use of product differentiation systems can be an effective way of managing the development of liabilities resulting from mislabelling of transgenic food products. Deciding what is driving the delivery of the food product to the marketplace is crucial for managing the labelling of the product in the marketplace. Identity-preserved production and marketing systems originally put in place to capture price premiums, segregation for food safety and traceability to identify product origins, if used properly, can greatly reduce the incidence of unintentional co-mingling and therefore, product mislabelling. Ensuring that the final food product is labelled as accurately as possible will lessen the opportunity for liability lawsuits to arise from improperly labelled food products.

Conclusions

Biotechnology innovations in agriculture present a clear challenge to the traditional marketing system. Transactions for new, proprietary, novel-trait crop varieties require a more extensive set of institutions than for traditional commodity varieties. Companies assisted by governments and industry associations have developed product differentiation systems that handle both the risks and assist with capturing the returns from the introduction of new products with commercially valuable input and output traits. Spot markets are increasingly competing with proprietary vertically integrated supply chains. The , optimal structure and organisation of these new supply chains has not evolved yet, but over time one would expect a more stable set of relationships to emerge.

More clearly articulated product differentiation offers biotechnology two immediate benefits. First, while it is evident that the use of product differentiation systems will be expensive to operate in the initial stages, these cost increases must be balanced against any expected or realised gains. Although research and seed development companies will have increased costs, they will gain significantly in terms of market adoption. As discussed in Chapter 8 it is very likely that if Monsanto and Aventis had introduced GM canola without IPPM systems, farmers would have shunned the new seeds. Second, product differentiation offers biotechnology firms another means to protect their investment into new technology crop developments. The problem with the current package of intellectual property rights is that they do not fully control the use of a new technology once it is expressed in seed. Most GM crops can be propagated in subsequent years with seed from previous years. While regulations and private contracts attempt to manage that activity, many in the industry note that they are far from effective.

The driver of each product differentiation system differs for important reasons. The use of IPPM systems is driven by private firm initiatives to capture the value associated with a special trait. Segregation systems are driven by regulatory agencies, where the objective is to prevent a potentially hazardous commodity from entering the supply chain for human food products. Traceability systems are driven by the supply chain to manage liabilities of product quality failures.

This review of existing product differentiation systems offers a number of lessons for both industry and government.

First, the examination of product differentiation systems has shown that the use of these systems is rising in agriculture. The increased use of these systems and the growing role of regulatory agencies in on-farm standards will continue. Essentially, more vertically integrated supply chains for products produced and marketed are developing. As an increasing variety of commodities are produced using product differentiation systems, producers will face a wide array of options that may complicate their farming operations. Industry and government may need to assist farmers to gain the expertise to properly evaluate their options.

Second, while we know that these systems are costly to develop and manage, it is not clear that the current estimates would reflect the costs of significantly expanded use of product differentiation. The cost figures for IPPM systems that have been documented range between 15% and 20% above the cost of conventional supply systems. It is crucial to remember that these IPPM systems operate at low volumes—usually less than 300,000 tonnes a year—and no system has been documented operating at a substantial volume, (e.g. one million tonnes). This has led some, both within the biotechnology industry and academics, to suggest that the cost of operating large product differentiation systems has been greatly under-estimated. If this is the case, then it is questionable whether large-scale product differentiation systems could achieve economies of scale that would support long-term use. This may have an impact on the introduction and commercialisation of future GM crops.

Third, products that require product differentiation systems must earn a return high enough to justify the expense. The absolute and relative costs will vary greatly depending on the agronomic traits of the crops, tolerances chosen and base value of the commodity crop. There are a number of examples of product differentiation systems that have been established that did not earn the necessary value in the marketplace, causing the systems to be abandoned. Thus far, all of these systems have been purpose built and rely solely on the capacity of their managers. There can and should be ways to reduce the costs of these independent systems by pooling certain activities and seeking scale economies.

Fourth, the cases highlight that to succeed the systems not only need to cover all of the costs, but the benefits need to be distributed so that no one in the chain is worse off than they would be outside the system. This is necessary both to keep participants involved in the system but also to reduce the risk of any cheating by participants that would undermine the system itself. With the approach of third generation GM crops that contain valuable output-based traits, product differentiation systems will become crucial. The StarLink™ maize example highlights that a product differentiation system is only as strong as its weakest link. All it takes is one participant or group becoming lax and allowing co-mingling to occur to not only destroy the system but also to inflict significantly larger costs on the marketplace. Studies suggest that while effective product differentiation systems may be costly to develop and run, they may well be cheaper than failure.

Finally, on-farm assurance of crops appears to be one potential area of weakness. Weak farmer compliance certainly contributed to the StarLink™ maize failure and are thought to have challenged the Monsanto and Aventis systems for HT canola. As more crops are contracted, there is significant potential for conflicting and confusing information to be disseminated to producers. There is ample room for industry and government to consider how they might assist in managing this aspect of product differentiation systems.

Part IV:

Prognosis

Chapter Ten:

Liability of Plant-Made Pharmaceuticals

Introduction

The introduction and rapid adoption of GM crops at the close of the 20th century was arguably one of the defining moments in the history of agriculture. As outlined in previous chapters, the rapid global adoption of GM varieties of canola, maize, cotton and soybeans was accompanied by considerable controversy. The contentious debates show no signs of abating. The array of issues is wide and varied, ranging from social issues such as consumer acceptance to scientific issues relating, for example, to gene flow.

One new issue has surfaced in recent years—how to regulate and manage the risk of pharmaceutical proteins in food plants. There is both great opportunity and great risk involved. On the one hand, the application of this technology has considerable potential to increase availability of pharmaceutical products to consumers. Estimates by the pharmaceutical industry suggest that as little as 1.5 acres of some pharmaceutical crops could satisfy the therapeutic needs of 10,000 patients. As the baby boomer population ages in developed countries, the demand for therapeutics will rise, raising the importance of researching new protein and antibody delivery systems. On the other hand, the use of food plants as the host organism for the expression of pharmaceutical proteins and antibodies is fraught with peril if the pharmaceutical plants are allowed to co-mingle with other food plants and enter the supply chain for human consumption. The detected presence of co-mingled pharmaceutical plants in processed food products has the potential to greatly disrupt (and destabilise) markets.

This chapter examines what may be the ultimate challenge for markets and regulators. It reviews the science and economics of pharmaceutical gene flow, discusses in detail the development of regulations targeted at controlling gene flow and reviews the results of commercialisation of these crop varieties. Up to this point in the book, we have examined issues where liabilities have been in existence and documented, now we move to discuss an issue that will be of great future importance. The intent of this forward looking analysis of potential areas of liability is to provide an illustrative example of where biotechnology is heading. This chapter identifies some contentious issues that will need to be addressed by regulators and industry stakeholders before such innovations will be allowed to reach their full potential.

The Pressure to Use PMPs

The cost of producing pharmaceuticals by conventional means is high, with the average new drug costing over US$260 million to develop and commercialise; some drug companies report spending in the range of US$700-800 million for specific new drugs. In an attempt to lower the cost of producing base components (proteins and antibodies) for new drugs, some firms are conducting research into using plants as vectors of expression.

Scientists and industry offer two justifications for pursuing plant made pharmaceuticals (PMPs). First, production of high-quality biological material is presently done using mammalian cells inside a bioreactor, which is very expensive and results in high drug costs that could potentially limit the number of people that benefit from new drugs. Second, there is an insufficient level of bioreactor capacity available to meet the current demand, let alone the expected increase in demand over the next decade.

The current prevalent technology that uses mammalian cells to produce human antibodies generates costs in the range of US$105-175 per gram. It has been estimated (McCloskey, 2002) that PMPs might be able to produce the same amount of antibodies at a cost of US$15-190 per gram. The range of variation in anticipated PMP costs arises from the prospect that the use of PMPs will lower production costs to a level that is economically feasible for potential new proteins that would, at present, be prohibitively expensive to produce. A number of countries have experimented with pharmaceutical crops, including the initiation of field trials, but only Canada and the United States (US) have any long-term history pertaining to the development of regulations targeted at pharmaceutical field trials and sustained experience with field trials.

In the final analysis, the issue comes down to cost. Table 10.1 compares the costs of the various antibody production systems. The production of these antibodies is valuable for the treatment of two types of diseases, including those which affect the general population, like arthritis, viral infections and cancer and those which affect smaller cohorts of people with particular inherited disease or metabolic disorders.

Furthermore, mammalian cell bioreactors take an average of five to seven years to build and cost on average US$600 million to complete. Given the lumpiness of this investment, there is a real chance that there could soon be inadequate supply. McCloskey (2002) estimates that given current trends, 20 to 50 products could be delayed due to the unavailability of bioreactor capacity. Currently, the production of four pharmaceutical products requiring biologics occupies 75% of mammalian cell fermentation capacity. By the end of this decade, there could be more than 80 competing antibody-dependent products with an estimated value exceeding US$20 billion, provided adequate production systems can be developed. The potential size of the market underlies the importance of exploring the potential of developing PMPs.

As the technical possibilities of, and demand for, human antibodies grow, there will be increasing pressure to use PMPs, forcing industry and government to consider the appropriate regulation of those plants. Of paramount importance will be assurances that the production of pharmaceutical proteins in food plants will not co-mingle with conventional crop production destined for human consumption. The detection of drug proteins in processed food products could threaten the social trust in pharmaceutical crop technology and ultimately destroy the ability to take advantage of this technology.

Table 10.1: Relative costs of various antibody production systems

	US$ M to produce 300 kg of antibodies
Maize	8
Eggs	14
Tobacco	18
Goats	57
Bioreactors	73

Source: Pew Initiative on Food and Biotechnology, 2002

Debates surrounding gene flow have grown in importance over the past several years, especially with the commencement of crop trials involving plants expressing pharmaceutical proteins. There is a growing level of concern that if field trials involving PMPs are not properly managed and controlled, that either the pollen or the seeds themselves will create food chain liabilities. Cross-pollination between PMP varieties and existing food crop varieties has the potential to create problems for the food processing industry. The problem expands if seeds from PMP field trials become co-mingled with raw products destined for human consumption, as processing firms would have to increase testing of the material and possibly undertake expensive product recalls.

Gene flow between transgenic plants and conventional plants and weedy relatives has been a hotly contested issue in recent years. New research is being undertaken that challenges conventional thinking about gene flow, but so far there is no consensus among the scientific community regarding gene flow. Debates have mainly focused on the ability of transgenic food crop varieties to cross-pollinate conventional varieties and whether the resulting progeny are biologically viable. A parallel debate is focusing on gene flow and progeny viability with weedy relatives of transgenic crops (see Chapter 7). While literature is emerging regarding transgenic gene flow, there is a noticeable absence of literature regarding cross-pollination and gene flow involving pharmaceutical plants.

Evolution and Value of PMPs

The first recombinant DNA proteins of human therapeutic value, interferon and insulin, were produced in bacteria. The justification, at the time they were developed in the mid 1970s, was that the fermentation and downstream processing technologies were mature, convenient and cost effective. Biotechnology techniques were next used in transgenic animals in 1980 and transgenic plants in 1983.

These techniques of biotechnology had multiple applications, many of which initially had little direct relationship to clinical medicine. Although initially promising, the production of human therapeutic proteins such as antibodies in mammalian cells or animals have not had a wide adoption. Antibodies produced in mammalian expression systems are expensive and difficult to scale up and pose safety concerns due to potential contamination with pathogenic organisms or oncogenic DNA sequences. Plants however, have become the principal focus for the production of antibodies, enzymes, vaccines and other therapeutic agents. With advances in transformation and expression systems it can be expected that more pharmaceutically important genes from many species will be inserted into plants. Table 10.2 provides a comparison of the existing recombinant human protein production systems.

As Table 10.2 shows, the potential economic benefit from using transgenic plants as the basis for expression of pharmaceuticals is high compared to other production vectors. The cost of using plants to produce pharmaceutical proteins is relatively low and some plants may be able to produce pharmaceutical proteins for an extended period of time. The quality of the product is expected to be high relative to some alternative production systems. The risk of contamination from other sources of contaminants is low and importantly, the cost of storing PMP products is expected to be relatively inexpensive.

Present PMP Production

Plant-made pharmaceuticals field trials began in 1992 in the US (Table 10.3). In North America, there was little need for pharmaceutical trials through the early to mid 1990s. In the US, trials became more common in 1998 and appeared to have declined beginning in 2001. Trials accelerated earlier in Canada (beginning in 1996), but peaked earlier at a lower level and declined following 2000. Other countries recorded a small number of PMP field trials between 1995 and 2002. There was no identifiable advantage for either country in the early stages of PMP crop research. Canada enjoyed an advantage from 1996-1998, while the US enjoyed the advantage from 1999-2001. Both countries converged again in 2002. Canada's early advantage was from canola research, which proved in the late 1990s to present too many safety concerns to proceed. The US lead was based upon research using maize as the plant of choice between 1999 and 2001. After problems in 2002, maize may be falling out of favour in the US for crop trials, while trials are climbing in Canada again based on testing with safflower.

Table 10.2: Comparison of production systems for recombinant human pharmaceutical proteins

	Bacteria	Yeast	Mammalian cell culture	Transgenic animals	Plant cell cultures	Transgenic plants
Overall cost	Low	Medium	High	High	Medium	Very low
Production timescale	Short	Medium	Long	Very long	Medium	Long
Scale-up capacity	High	High	Very low	Low	Medium	Very high
Product quality	Low	Medium	Very high	Very high	High	High
Glycosylation	None	Incorrect	Correct	Correct	Minor differences	Minor differences
Contamination risks	Endotoxins	Low risk	Viruses, prions and oncogenic DNA	Viruses, prions and oncogenic DNA	Low risk	Low risk
Storage cost	Moderate	Moderate	Expensive	Expensive	Moderate	Inexpensive

Source: Ma et al., 2003.

Table 10.3: Field trials of plant-made pharmaceuticals, by country

Country	92-95	96	97	98	99	00	01	02	Total
Total	5	12	8	22	23	27	23	14	134
US	1	1	1	8	12	14	18	7	62
Canada	3	11	5	11	8	6	3	6	53
Japan	-	-	-	-	-	7	-	-	7
France	1	-	1	-	3	-	1	-	6
Argentina	-	-	-	3	-	-	-	-	3
Australia	-	-	-	-	-	-	-	1	1
Italy	-	-	-	-	-	-	1	-	1
Spain	-	-	1	-	-	-	-	-	1

Source: CFIA, 2003; APHIS, 2003; OECD, 2003; CONABIA, 2003;
European Commission, 2003; OGTR, 2003; and Japanese MAFF, 2003

There are several reasons for variations in the number of field trials during the past decade. First, the issuance of approvals differed amongst nations and regions. Second, the pace of discovery can be serendipitous or planned. The crop pharmaceutical industry itself is in the early stages of development so that there is great uncertainty and uneven demand in their need for trials. As such, field trials, as with the conventional pharmaceuticals clinical trials, are unlikely to have a predictable trend. Third, variations in field trials can occur because of seasonal and/or environmental conditions that dictate postponement of trials. Fourth, as judged by the analysis of the types of crops, fluctuations in the numbers of pharmaceuticals that could be derived from any one crop should be expected. The genetic and physiological constraints in plants place limits to their use for transgenic plant construction, both in the food and pharmaceutical contexts (Khachatourians *et al.*, 2002).

The evolution of PMPs has mirrored the research trends in agricultural biotechnology: transformation research started with tobacco, moved to dicotyledons like canola and finally to monocotyledons, where the first research was with rice. The different crop varieties used for pharmaceutical trials are shown in Table 10.4. Canola was an early favourite due to the amount of canola transformation research that had already taken place and its attractive oil properties. After several years of pharmaceutical crop experimentation in Canada, it became obvious that canola was not a suitable host plant due to the high incidence of pollen flow and the threat posed to the large commercial canola growing industry in Western Canada. At the time that pharmaceutical canola trials were ending, trials started with maize, rice and tobacco. The use of maize and tobacco for pharmaceutical trials grew between 1997 and 2001, while experiments in rice have been minimal.

Table 10.4: Field trials of plant-made pharmaceuticals, by crop

Crop	92-95	96	97	98	99	00	01	02	Total
Alfalfa (Lucerne in Europe)	1	-	-	-	-	-	-	-	1
Barley	-	-	-	-	--	-	1	-	1
Canola	3	2	8	4	-	-	-	-	17
Clover	-	-	-	-	-	-	1	-	1
Maize	-	-	1	5	11	11	14	4	46
Flax	-	-	-	1	1	2	-	-	4
Mustard	-	1	-	2	-	-	-	-	3
Poppy	-	-	-	-	-	-	-	1	1
Rice	-	-	1	2	2	2	2	-	9
Safflower	-	-	-	-	-	1	2	4	7
Sugar Cane	-	-	-	-	-	-	1	-	1
Tobacco	1	-	1	-	5	5	9	4	25
Tomato	-	-	-	1	-	-	-	-	1

Source: CFIA, 2003; APHIS, 2003; OECD, 2003; CONABIA, 2003; European Commission, 2003; OGTR, 2003; and Japanese MAFF, 2003

Experiments with the use of flax occurred briefly in Canada in the late 1990s, but concern about seed dispersal during harvesting limited this research. Trials in safflower have commenced in the past three years as this research has replaced the previous canola research. A variety of other crops such as forages, vegetables and flowers have been experimented with, but it would seem that little in the way of useful pharmaceutical potential is currently available in these plant varieties. The one exception to this may be the use of poppies in Australia for improved production of opium. Field trials of transgenic pharmaceutical poppies started in 2002 and it would appear that there is some long-term potential utilising poppies.

There was a noticeable decline in pharmaceutical crop trials in 2002. There may be three reasons for this. First, crop trials are cyclical by their very nature and this may be nothing more than a natural dip in the number of trials. Second, pharmaceutical companies have been conducting trials for 5-6 years with some crops and have now completed Phase 3 clinical trials and are waiting to see what the financial outcome will be prior to commencing new research. Third, pharmaceutical companies want to control the pipeline for this technology. By limiting the number of trials, they may be able to restrict supply and thereby increase their profits on each candidate plant.

Table 10.5: Transgenic crops and human consumption

Crop category	Specific transgenic crops (either approved or in trials)	Use in pharma ceutical trials	Cross- pollination potential	Modality of consumption	
				Plant tissue(s) and organs*	Extracellular plant metabolic ingredients
Cereals	Maize, barley, rice, wheat	Yes	Low- Medium	Direct	No
Oilseeds	Canola, flax, mustard, cotton, safflower	Yes	High	Mainly indirect	Yes
Pulses	Soybean	Yes	Low	Direct and indirect	No
Forages	Alfalfa, clover, tobacco, sugar cane, sugar beets	Yes	Medium	Very minimal indirect	No
Fruits and Vegetables (including juices)	Poppy, cantaloupe, melon, radish, potato, squash, tomato, banana, strawberry, lettuce and papaya	Yes	Low-High	Direct	Yes

Note: Includes plant cells, tissues and organs that include rDNA and/or primary or secondary metabolites (i.e. excluding DNA such as oils, starches, proteins, amino acids and processed materials and tissues, including juice)

Source: Authors.

The three leading crop species used for pharmaceutical trials by design (availability of vectors and transformation systems) and/or choice (agronomy and growers) have been maize, tobacco and canola. The problem is that maize and canola are intended for human consumption and, the potential for co-mingling or cross-pollination exists, raising concerns about using these species for PMP trials. Table 10.5 presents the major transgenic crops, identifies whether they are used in PMP trials and examines their modality of consumption.

Regulation of PMPs

The challenge of regulating pharmaceutical crops is with the overlap of medicine and agriculture. Drug companies are beginning to use plants as expression vectors for proteins and antibodies that are used in the production of new drugs. The regulation of pharmaceutical crops is not fully formalised, as it involves in the US the USDA, the EPA, the FDA and, at times, the National Institutes of Health (NIH). All have a role and are involved. In Canada, the regulation of PMPs involves Health Canada, the CFIA, Environment Canada and the Canadian Institute of Health Research. A clearer model of regulation is required to adequately address the emerging liabilities from pharmaceutical crop regulatory violations.

Ultimately, regulatory systems are designed to assess risks of new products or processes, using a scientific risk assessment framework. When a firm violates the regulations or the regulations fail to properly assess and manage the risks, a liability is triggered. As pointed out in earlier chapters, negative media coverage can result in some consumers that initially are indifferent to a specific innovation becoming concerned about, or even opposed, to the innovation and possibly to all innovations in general. The bottom line can be a loss of social trust in innovation, albeit specific or general. This is not to say that the decline in trust is long term, or that it can not be reversed, but rather that there will be fewer consumers willing to express support for an innovation or the resulting commercial products.

One key concept of liability that is particularly relevant to PMPs is the concept of strict liability. Strict liabilities are found for one-time, unnatural occurrences. With strict liability, the prosecution has to prove there was unnatural use of a product and that the plaintiff suffered harm, but the prosecution does not have to prove how the harm arose. As noted in chapter 2, the case of Rylands v. Fletcher (1868) is frequently cited in legal literature as a reference for strict liability. In this case, Fletcher owned land that he had surrounded with a dam and filled the land with salt water. Rylands owned a mine near by and when the dam gave way, it flooded the mine. The court ruled that the liability lay with Fletcher, as it was unnatural to have a reservoir of salt water contained in that area. The ruling in Rylands v. Fletcher describes a situation where the product being stored was not naturally occurring and, therefore, the product was inherently dangerous.

The ruling in Rylands v. Fletcher has three important implications for the issue of gene flow in pharmaceutical crops. First, the drift of transgenic pollen is not a one-time occurrence, rather it happens annually, for a period of up to three to six weeks in many crop varieties. Second, it would seem implausible to argue that transgenic pollen is stored in any form or fashion upon a farm or a field test plot. The third consideration important to this issue is that the presence of pharmaceutical genes in crops destined for human consumption could be inherently dangerous. The key argument from Rylands v. Fletcher was that the danger was not naturally occurring. There is a strong argument to be made that PMPs are not naturally occurring and that the unintended gene flow from these crops may be foreseeably inherently dangerous.

The different forms of liability can be classified according to the following governance mechanisms. Criminal liability is strictly a legal issue and dealt with by the courts. In these cases, individuals who have broken the law are either punished financially or by serving time in a penal institution. Civil liability is an economic issue and, while it can be handled by the court system, it is more common for these cases to be settled out of court. Many civil lawsuits in the US are class action suits against large firms and courts often award financial compensation to those negatively affected regardless of the company's ability to pay. Executives of the offending firms are not normally prosecuted through criminal proceedings or even held liable for any court findings, unless there is demonstrated breach of their fiduciary trust. As argued earlier in this book, socio-economic liability, in contrast, is not dealt with by the court system, but rather, is reflected in the attitudes of the consumers within a given society. Whereas the other two forms of liability have recognised mechanisms to deal with the liablility, this form of liability incurs public costs reflected in the loss of trust in a product line or producing region, rather than simply directed at a specific company or a branded product.

The relationship between the stakeholders can be analysed using the inter-disciplinary institutional model outlined in Chapter 1. This model is again used to examine the central domain of each stakeholder and highlights the strengths of each stakeholder within their specific domain. Additionally, the model explores the interaction between the stakeholders in the areas of overlap, which are the crucial areas for fostering the development of trust. The model is useful in examining the development of regulations, industry standards and codes of practice. One implication of using this framework is that it highlights the various systems of governance that are designed, and intended, to work in harmony with each other, and provides insights into those times when there are gaps or oversights within the governance structure that result in regulatory failures.

Thus, each sector has control over a central domain, which can be described as their portion of the sphere that does not overlap with any other sphere. In the areas of overlap, however, the jurisdiction and incentives can be difficult to discern. In the US, regulations for biotechnology are the shared responsibility of the USDA, the EPA and the FDA (domain A in Figure 1.1). The combined regulations of these government agencies have, for the most part, been effective at preventing regulatory oversights. The private firms (B in Figure 1.1) in the agriculture biotechnology industry have developed their own operating procedures, which enact and manage the regulatory decisions and in some instances have greater stringency than the regulations established by the federal regulatory bodies. Industry associations (C in Figure 1.1) in the US, such as the Biotechnology Industry Organization (BIO), have worked progressively with the private firms to develop industry standards (a collective activity that requires demonstrate to the public that the industry is conscious of potential concerns and have taken action to try and ensure that no oversights exist.

The response of regulatory agencies has differed between Canada and the US. Canada has adopted a three channelled product differentiation system that distinguishes products based on economic criteria. The analysis in Chapter 9 demonstrated how identity-preserved production and marketing systems are used by private firms to capture premiums for niche market products, how the CFIA has implemented segregation of industrial crops (such as high erucic acid rapeseed) that could endanger food safety through co-mingling and how retailers and others in the supply chain are implementing traceability systems to meet consumers demands for more timely product recalls and tracing. This array of product differentiation systems has worked well in Canada, while in the US, the difference is that the federal regulatory agencies involved in the regulation of transgenic crops are less specific about the purposes of their systems—for example they have never demanded the same level of segregation of crops as in Canada.

Recently, the federal regulatory bodies in the US have relaxed some of the regulations relating to crop production and a process is in place for applications to deregulate some transgenic crop varieties. The federal regulatory agencies in the model have moved out of the portion of the spheres where overlap occurs, but this regulatory withdrawal has not been followed-up by actions initiated by the private firms or the industry associations. The lack of progressive actions from private firms and the industry associations has resulted in the creation of a regulatory gap for the field testing of PMPs. The result of this was a series of regulatory violations between 2001 and 2003, which prompted the federal regulatory agencies in the US to respond by strengthening the regulation of PMP crop trials.

The PMP actors in the areas of overlap in the institutional model have yet to be identified. In this emerging sector of research, the science is well ahead of the regulators capacity to regulate and society's appreciation of the innovation.

Strong governance institutions are especially important for the production of PMPs, which have the possibility of entering and endangering the human food supply chain. These governance institutions currently range from national regulatory agencies, to private industry organisations, to judicial systems. An international comparison of the three leading forms of governing institutions (Table 10.6) illustrates which institutions lead in different markets. The commercialisation of PMPs varies greatly from country to country, depending upon how far the actual governance system diverges from a comprehensive regulatory regime—what we posit would be the optimum.

The optimum would include three strong institutional pillars that are able to anticipate and manage risks. This would require a strong regulatory body that anticipates public good issues of concern to society as a whole and begins to develop regulations prior to the commercialisation of products. Strong industry and civil society associations are also needed to operate as progressive lobby groups with a wide network of industry and social representation that can develop industry standards that either can become the base for regulations or can exceed the regulations provided by government. Finally, strong judicial systems exceed the regulations provided by government. Finally, strong judicial systems are needed to mediate issues relating to the commercialisation of transgenic crops and the ownership of the corresponding intellectual property (in effect, they keep industry operating and accountable).

Table 10.6: Assessment of national governance institutions for PMPs

	Type of governance institution		
	National regulatory body	Industry association	Judicial system
Canada	Strong	Medium	Strong
United States	Medium	Strong	Strong
France	Strong	Weak	Medium-Strong
Australia	Strong	Medium	Strong
Argentina	Weak	Weak	Weak
Japan	Strong	Weak	Strong
Italy	Weak	Weak	Not available
Spain	Medium	Weak	Not available

Source: Authors

A closer examination of the regulatory systems in Canada and the US reveal some surprising differences. Some would argue that the Canadian regulatory agencies have been more vigilant regarding transgenic crops than their US counterparts. Beginning in the early 1990s Canadian regulators stated that all transgenic crops (as well as many mutagenic crops) would be treated as plants with novel traits (PNTs) and, therefore, would be subject to additional mandatory regulatory oversight compared to conventional crops varieties. Every new PNT requires mandatory oversight of their trials and review of their efficacy and impact on safety of food, feed and the environment. Government agencies demand to see both the raw data and summaries of all tests performed and have the final say on every introduction. The Canadian system also has a formal system of contract registration for risky industrial crops and imposes criminal penalties for infractions. While the Canadian regulators have not completed their development of special rules for PMPs, they have been very influential in directing companies away from areas deemed to be of higher risk (e.g. canola) by simply reminding the developers that such products will not be approved. Meanwhile, the Council for Biotechnology Information (CBI) is a smaller association than in its counterpart in the US and has not developed the synergy that BIO enjoys in the US. At least part of the reason lies in the concentration of power in the Canadian government (through the Prime Ministerial structure, cabinet secrecy and party solidarity) that limits the association's ability to gain access or find supporters within the government apparatus. While the Canadian judicial systems is viewed highly in terms of its independence and professionalism, it is inherently weaker than in the US because of the limited use of class action suits and the very narrow parameters applied for punitive damages.

The initial regulations in the early 1990s in the US were viewed by the industry as being too lax and therefore insufficient to establish trust with consumers. In response, the industry asked the regulators to strengthen the regulations for transgenic crops. Nevertheless, the US regulatory system has consistently been less rigorous in the approach to dealing with transgenic crops than regulators in Canada—e.g. most reviews are voluntary, non-transgenic novel traits are not reviewed and the regulatory agencies only see study summaries rather than raw data. As in Canada, the US regulators have not sorted out how to handle PMPs. The extra challenge they face is that they do not have the same powers and legal authority that Canadian regulators have to direct developers away from crops. While the regulatory mechanism may be weaker, the other two domains are stronger. The industry association is considerably larger than in Canada and, given the more open nature of US governance, has better access and a stronger voice in the US than in Canada. BIO is viewed by many as a very authoritative voice when speaking on issues affecting the industry. The courts, similarly, are more engaged, partly because they are more open to class actions and because they award much higher punitive damages than in Canada. For instance, Aventis was pursued by a class action suit in the US claiming that the impacts of StarLink™ had depressed maize prices in the US and resulted in economic losses for maize producers. Faced with a potentially larger judgement, both parties settled very early into the trial, agreeing on US$110 million in compensation. On 6 March 2003, APHIS announced that they would strengthen permit conditions for field testing transgenic crops, including field trials for PMPs. The number of site inspections will increase to five during the trial and two the following season. The permits for pharmaceutical trials will state that no maize can be grown within one mile of the trial site and that no food or feed crop can be grown on the site the following season. The size of the buffer zone was doubled from 25 to 50 feet. This strengthening of the regulatory requirements, in part, can be seen as an effort to address the concerns that arose following the regulatory violations between 2001 and 2003.

With the exception of Canada and the US, there have been very few pharmaceutical crop trials and this creates a challenge when trying to evaluate the relative strength of the related governance institutions. Three European nations have varying levels of government regulatory oversight. France has been strongly opposed to transgenic crops and developed strict regulations for transgenic field trials, Italy has changed positions over the past five years and transgenic crops are presently forbidden, while Spain has annually averaged between 45,000 and 55,000 acres of *Bt* maize for the past five years (Brookes, 2002). While this is a relatively small amount of production, it does indicate that the Spanish regulators have developed a functioning regulatory system for the co-existence of transgenic and conventional cropping. The main industry association in Europe, EuropaBio, is a loose coalition of biotechnology firms operating in Europe, but due to the high level of organised opposition, the diverse nature of the EU and widely dispersed power and authorities in the EU, its voice is not heard loudly. The French judicial system has, albeit with a limited number of cases, protected the integrity of research and field trials of transgenic crops (ensuring the isolation of trials, even from protesters, is a foundational requirement for any effective regulatory regime), while the court systems in Italy and Spain have not been tested.

Australia, Argentina and Japan have allowed pharmaceutical trials to take place, but on a very limited basis. Australia's regulatory agency is modelled to some degree on those of North America and, therefore, has adopted a consistent policy for transgenic crops. The recent economic difficulties in Argentina have resulted in, at best, chaotic regulations. Japan has a very strong regulatory agency whose decisions are consistent with North American decisions, but lag by a period of several years. Australia has a developing industry association, but it is limited as Australia is just in the initial process of granting commercialisation to transgenic crops. Argentina and Japan have virtually no effective industry associations. Australia has a judicial system similar to that of North America but the federal constitution empowers each Australian state individually to approve or ban transgenic crops, which may possibly create a legal jurisdictional battle, with a number of expected lawsuits against the states enacting moratoriums. Again, the disruption of Argentina's economy has reduced the ability of its judiciary to provide consistent decisions. Japan's judicial system has historically been a strong supporter of biotechnology, but there is growing social concern about biotechnology and this may be reflected in future court decisions. In any case, the judicial system with its adversarial nature runs counter to the Japanese cultural preference for finding compromise through negotiation and trust building and, hence, is called upon much less often than in western jurisdictions.

Based on an analysis of the US and Canadian governance institutions relating to biotechnology, it can be argued that a functioning regulatory system requires strong institutions in all three pillars. Australia is developing a functioning regulatory structure, but only after careful observation of events in North America. All the other countries, Argentina, France, Italy, Japan and Spain, lack a strong institution in at least one of the three core domains. This lack of institutional leadership results in an imbalance of authority, which may indicate that either the government agencies have too much regulatory power and are unrealistic in their expectations of biotechnology companies, or that there is no structured bureaucracy capable of making consistent policy decisions.

Regulation Problems

As the technology of PMPs rapidly moves from laboratory to field, the regulations developed to control these new crop varieties have been severely tested. While regulators in the US have argued that the detection of ProdiGene's experimental pharmaceutical maize in a silo of soybeans late in 2002 is proof that the regulations are working, the simple fact that a pharmaceutical crop that was supposed to be contained on-farm actually reached a grain terminal without being detected, shows that the regulations are probably not sufficiently stringent. The containment of living plants is proving to be increasingly challenging given our inability to completely control nature.

The issue of gene flow in canola was documented by Smyth *et al.* (2002) and that situation remains unchanged. Scientists and regulators are still in a conundrum at best, or conflict at worst, about the impacts and regulation of gene flow. The issue of unintended gene flow first became a global news issue in the autunm of 2001 with the discovery that some varieties of Mexican maize contained transgenic material that should not have been there (Quist and Chapela, 2001). While this research was contested within the scientific community and is presently the subject of a North American Free Trade Agreement (NAFTA) Commission for Environmental Co-operation (CEC) Chapter 13 panel review, the concern continued into the summer of 2002 as a research team led by Allison Snow of Ohio State University reported preliminary evidence suggesting that a trait from transgene insertions in sunflowers may be able to move to other plants, thus creating the conditions for 'superweeds' (Snow *et al.*, 2003).

The first, and most widely publicised regulatory violation, was the case of StarLink™ maize which had been introduced by Aventis (Table 10.7). Aventis received split-run approval for this variety of maize, so that it could be produced within an identity preservation system for use as animal feed but not for human consumption. Although Aventis paid a premium of US$0.25/bushel to contain the maize and suggested rules to ensure it was used only for animal feed, the evidence suggests that Aventis and the EPA in the US did not do enough to ensure that producers were aware of the split-run approval. As a result, StarLink™ maize co-mingled with maize destined for human consumption. Close to 300 food products containing StarLink™ maize were detected and recalled, at an estimated cost of US$100 million. Aventis recently settled a lawsuit with affected maize producers for US$110 million.

The extent of the problem was more clearly defined in the autumn of 2002 and the spring of 2003 with a few high profile regulatory actions. In November 2002, ProdiGene Inc. was fined US$250,000 for allowing experimental pharmaceutical maize grown in 2001 to volunteer and grow to maturity within a soybean crop grown in the same field in 2002. The regulatory infringement was discovered by inspectors with USDA APHIS. The affected soybean crop was harvested and pooled in a commercial grain silo, thus contaminating an estimated 500,000 bushels of soybeans. The cost to ProdiGene for buying the contaminated soybeans and having them transported to be destroyed was estimated to be US$3.5 million.

The following month, the issue was once again making news headlines in North America. In mid December, Dow AgroSciences and Pioneer Hi-Bred were fined by the EPA for two separate regulatory violations. Dow was fined US$8,800 for failing to meet all the defined conditions to prevent gene transfer with an experimental transgenic maize variety undergoing field trials at

Molokai, Hawaii. The plot, which was 0.1 acres in size, failed to meet the EPA permit conditions because there was no windbreak of wiliwili trees in place and the bordering rows (outside 12-24 rows) were not of the variety specified in the permit. Pioneer was fined US$9,900 for an experimental transgenic maize variety in Kauai that was planted in an unapproved location that turned out to be too close to other experimental maize varieties. The Pioneer permit from the EPA specified an isolation distance of 1,260 feet, which was not observed.

Table 10.7: Transgenic crop regulatory violations, 1998-03

Company and crop year of violation	Crop	Location	Violation	Impact
Aventis CropScience, 1998-2000	Maize	Numerous states	Failed to prevent maize approved for animal feed from entering maize destined for human consumption	Paid US$0.25/bushel premium to contain maize, recall of nearly 300 maize food products at an estimated cost of US$100 million and settled lawsuit for US$110 million
ProdiGene, 2001-02	Maize	Nebraska	Volunteer maize growing in soybean field	Fined US$250,000 and forced to pay clean-up costs of US$3.5 million
Dow AgroSciences, 2002	Maize	Hawaii	No tree windbreak and bordering rows	Fined US$8,800
Pioneer Hi-Bred, 2002	Maize	Hawaii	Plot planted in unapproved location	Fined US$9,900
Dow AgroSciences, 2003	Maize	Hawaii	Plants detected with unapproved gene and failure to notify the EPA	Fined US$72,000

Source: http://131.104.232.9/agnet-archives.htm.

Finally in April 2003, Dow was again fined for violating an EPA permit in Kauai. This time the fine was US$72,000 and resulted from the detection of 12 transgenic maize plants that contained an unapproved gene that is suspected of coming from the pollen from another experimental plot located nearby. Although Dow officials discovered this unplanned gene flow, Dow failed to notify the EPA promptly and EPA officials expressed disappointment over that delay. The report in the *Washington Post* (April 24, 2003) argued that this incident was "... the latest setback for a biotechnology industry struggling to comply with government rules. ... some advocates say the problems cast doubt on a fundamental premise of government policy: that experimental varieties of corn or other crops can be planted in fields but kept out of food crops."

Four separate, but related, regulatory violations within a six-month period may be nothing more than a freak occurrence and may never happen again. What is troubling, however, and more representative of the real issue, is that these regulatory violations reflect a lack of commitment and understanding of the importance of a transparent, accountable and effective regime for new trait crops.

Conclusions

The challenge of PMPs is going to be to develop a structure for a fully integrated regulatory system that effectively evaluates, manages and communicates about the risks of a system, and ultimately one that both enforces and is seen to enforce failures. In spite of the US regulatory changes, there is an apparent inability of regulators to enforce the regulations. In the ProdiGene case, the cost of the fine, clean-up and destroying the contaminated soybeans was estimated to be US$3.75 million. The problem with imposing such a large fine on a small biotechnology company is that there is seldom enough cash-flow within the company to pay a fine of this magnitude. In this case, the US government had to lend ProdiGene the money to pay the fine. This is symptomatic of the bio-pharma industry as a whole, as small bio-pharma companies often do not have sufficient financial resources to pay large regulatory violation fines. The problem is that if firms know that governments will provide loans or loan guarantees in the event of fines from regulatory violation, what incentives exist for the firms to adopt standards that improve the control of pharmaceutical crops? If existing enforcement mechanisms are found wanting, or are lacking, can trust be sustained?

While the model illustrates a useful starting point, these models are not in equilibrium in many of the countries undertaking PMP crop trials. In the US, it can be argued that none of the spheres are overlapping with each other. This creates regulatory gaps that, little by little, erode society's faith in the ability of government and industry to manage these new crop technologies. The intention of the crop trial process is to build integrity for the crop variety participating in the trial process. When regulatory violations occur within this process, not only is the integrity of the process diminished, but the merit of PMPs being grown in food crops is questioned.

The challenge would seem to be that in countries where the regulators are unwilling or unable to step forward and be the leading and dominant institution, private industry is shirking the responsibility as proper guardians of new innovations. Similarly, where industry organisations are unable to generate consensus on, or adherence to, proper standards and procedures, governments have often been unwilling or unable to fill these gaps. One option is to let the legal system step in and establish standards and regulations based on decisions from multiple lawsuits. This may well occur if the major stakeholders do not begin to take their responsibility to society regarding the production of PMPs more seriously, but it is unlikely to occur quickly or at low cost.

Chapter Eleven:

Handling Liabilities from Transformative Technologies

Introduction

Transformative technologies create both change and uncertainty, which makes dealing with any resulting risks and liabilities much more complex. Incremental technological change, where potential impacts are relatively narrowly confined and predictable, tends to be easier to manage because the potential uses and users are known, a restricted and well defined group of stakeholders are usually engaged from the start and the adaptation, adoption and diffusion are limited to defined products or markets. In contrast, transformative technologies are by their very nature unpredictable. In the past, a variety of transformative technologies have radically altered the world, usually in unexpected ways. The development of electricity provided a new and more flexible form of power, which when combined with electric motors and electric filaments, fundamentally altered our organisation of business and way of life. More recently, information, communications and telecommunications (ICT) technologies have radically changed how we develop, store, communicate and use knowledge and information. These new technologies have altered our working and home lives, but have also fundamentally altered geo-politics, as governments can no longer easily control access to competing sources of information.

Biotechnology, coming on the heels of the ICT revolution, shows all the signs of being one of the greatest and, perhaps, most disruptive transformative technologies to come along in modern times. The potential impact and extent of this technology is unprecedented. More than 40% of our global economy is engaged in bio-based activities and innovation based on biotechnology may provide new competitive alternatives for much of the rest of our inorganic industrial activity. Further, biotechnology could ultimately allow us to disconnect our food and fibre production from the unpredictable and variable natural environment, while at the same time cause us to reconsider how little or much we want to craft our own destiny as 'creators' of life. The scale, scope and speed of these changes create profound uncertainties and ultimately engender new liabilities.

The Challenge of Managing Liabilities

This book has presented the argument that, at least in the area of agricultural biotechnology, the uncertainties and resulting legal, commercial and socio-economic liabilities must be recognised and addressed in a forthright manner. The challenge is to find a way to do that.

In the first instance, it is becoming increasingly difficult to establish what the science is, where it is going and how it might be applied safely and efficaciously. The traditional peer review system of managing the normalisation of emerging science has not kept pace with the ICT revolution. Part of the challenge is the increasingly private nature of research, with more than 70% of biotechnology research undertaken either by private sector scientists or by public sector and university scientists funded by, and working collaboratively with, private interests. The goal of science is no longer simply discovery of new knowledge; increasingly scientists want broader recognition and commercial gain. In the past new scientific developments were tried, tested and challenged in a slow, methodical way outside the limelight. Now, each new discovery (touting either new unimaginable benefits or unfathomable risks) is put in front of a largely scientifically unsophisticated public with great fanfare through glowing press releases to traditional media outlets and on the Internet, hyped lead articles in learned journals, perfunctory front page stories in tabloids and quality broadsheets and in brief, sensationalised sound bites on radio and television. A circus-like atmosphere of debate characterised by excitement and fear has replaced the traditional period of contemplation, assessment and testing (Caulfield, 2002 characterises this as 'genohype'). Thus, the normal touchstone of sage advice from learned scientists is no longer available.

At the other end of the innovation process, citizens and consumers are often confused, angry and uncertain about what to believe, who to trust and what to do. While the average citizen in North America, Europe and most OECD countries may be exponentially more knowledgeable and worldly than their ancestors, most still are unable to absorb the basic concepts pertaining to this technology. Their uncertainty is compounded by the central role food plays in any society. Food is both simple—it is what we eat to sustain life—and extraordinarily complex—what we eat includes whole and processed plants, animals and fish from all over the world, produced in a bewildering variety of ways (Phillips and Wolfe, 2001). Further, any agricultural production alters the environment. In recent years any change that has the potential to alter the environment has come under increased scrutiny from concerned members of civil society. The evidence so far suggests that neither consumers nor citizens are clear about what they want from agricultural biotechnology. In the absence of clear signals from consumers, food presents unique governance challenges.

The institutions that have been developed over the past 100 years to handle risks and adjudicate liabilities in the food system are increasingly strained to handle the uncertainties created by these new products.

Regulatory systems, for example, were all designed before the advent of biotechnology and are based on principles and procedures that underpin all government operations. The underlying principles of regulation (transparency, accountability, non-discrimination and responsibility) and the undertaking of scientific risk assessment (SRA) have become the foundation for all risk analysis and are now firmly entrenched in all countries and all international institutions. That poses two major problems. In the first instance, SRA systems have as their basis characterisations of hazards and exposures that require a base of scientific theory that provides causal explanations for hazards, a set of methodological approaches that can garner results and a body of evidence that can provide assessments of probabilistic outcomes. In the past, Kuhn's paradigmatic 'normal science' provided the requisite theory, methods and evidence (Kuhn, 1970). With the declining efficacy of peer review processes, regulators are now faced with addressing issues arising from emerging science. As a result, many regulatory agencies are developing codes of conduct (often called a 'precautionary principle' or 'precautionary approach') to address how they will deal with science that offers preliminary cautionary evidence but does not meet the standards of 'normal' science. Thus far there is no norm for how to handle this incomplete evidence.

Regulators are also keeping an eye on what their consumers or citizens will tolerate. The difficulty is that most of the methods for characterising public concerns are not definitive. Most of the surveys, focus groups, citizens' juries, deliberative consultations and interactive experiments provide incomplete and often contradictory results. Many of the methods for obtaining input from the public are open to undue influence, or capture, by those in society that have strong preferences regarding a particular issue. Even when they provide clear views (e.g. the desire for mandatory labelling), both the structure of the questions and the nature of the answers are for the most part one-dimensional, offering only a yes/no statement of opinion, and do not probe the intensity of opinions and the range of real-world trade-offs that might constrain options. As a result, even definitive results are challenged and debated vociferously by those with other opinions or interests. Furthermore, consumers are not simply concerned about the health and safety aspects of agricultural biotechnology. They also express concerns about the distribution of economic benefits and costs and the ethical and moral aspects of its use. None of the current regulatory systems is designed to address these issues—their primary and, for the most part, only purpose is to assess health and environmental risk. Thus, governments find that they do not have mechanisms to handle socio-economic issues. Ultimately, political systems are most able and willing to act decisively when there is either a clear, empirically-based imperative (e.g. definitive science) or when there is overwhelming public pressure for change and an unambiguously appropriate and popular policy option. Transformative technologies provide neither.

In the absence of clear information and signals from science and civil society and faced with inconclusive or contradictory messages from the regulatory and policy systems of government, industry is forced to become more engaged. Both individual firms and sectoral associations have significant resources invested—either in the existing products and their supply chains or in new proprietary technologies and products—which need careful management to sustain or exploit. Both firms and associations have been active in trying to find some middle ground that allows co-existence of competing technologies, so that new investments can be commercialised and pre-existing markets can be sustained. While there have been some successes, the evidence reviewed above suggests that there is often a mismatch of incentives, authorities and interests in those systems that can lead to failures, which create socio-economic and commercial liabilities.

Thus far, the situation has been worsening and not improving. Science is becoming more fractious, consumers and citizens more jaded, regulators more frustrated and industry more anxious. If the technology is going to provide future benefits, the issue of liability will need to be addressed.

Quick Fixes for Liabilities from Transformative Technologies

As the pace of life accelerates, people often look for quick fixes to problems that emerge. The area of agricultural biotechnology is no exception. If you were to ask any of the many individuals or groups involved in the debate about biotechnology, they would all have their favourite 'quick fix' or 'silver bullet' to solve the problem. It is instructive to review their perspectives.

Scientists instinctively ask for more resources—time, money, materials, equipment, and infrastructure—to do more science. While scientists can provide some new answers, they cannot be expected to resolve all of the issues related to a transformative technology. In the first instance, scientists are trained to ask a specific set of questions and apply specific tools in their investigations. Given the nature of transformative technologies—that is that all of the potential uses and impacts are unknowable because of the widespread application of the technology—it is unlikely that scientists unaided would be able to frame the appropriate questions to guide experiments. Whatever science is done will need to be informed by the needs, desires, constraints and opportunities of consumers, citizens, regulators and industry. Rather than more resources, scientists may need to try new interdisciplinary models of investigation.

At the other end of the scale, many non-governmental organisations argue that the paucity of scientific certainty is too threatening to allow either any further use of the technology or commercialisation of products developed with the technology. Essentially, they are calling for a pause, either in the form of a ban or other prohibitions, while we seek out greater certainty. The challenge with this approach is that the ubiquitous nature of transformative technology means that the risks, and benefits, can only be found by exploring it further. Many argue that most of the risks that create liabilities will not be those that are anticipated in normal use, but rather will be unanticipated impacts caused by 'off-label' use of the technology and its products. Some argue that this justifies a ban, while others suggest this justifies a clearly delineated set of rules, responsibilities and obligations that mobilises a more vigilant and engaged industry, government and citizenry.

Sometimes it appears as if consumers have lost faith in almost everyone except themselves. Rather than sort out who is right, many consumers would simply like to be able to make their own decisions about whether and how to use the products of this transformative technology. To that end, surveys suggest consumers almost universally want mandatory labelling of products of biotechnology. The difficulty is that none of the requisite actors are positioned to offer credible, affordable, appropriate labelling of their foods, whether for GM traits or for other experience or credence attributes. This demand for more information is almost unprecedented.

Government regulators are somewhat confused about what to do. While all governments take their role seriously as regulators of public health and environmental safety, they appear at times to be troubled by their other roles as developers and promoters of new products. Furthermore, virtually every national regulatory system is dependent on others—scientists and their professional associations, firms, other national regulators and international regulatory authorities—to successfully and economically undertake accurate risk assessments, manage any attendant risks and communicate to their citizens and consumers about the nature of any product risks. Ultimately, more or new regulations could be part of a solution, but it is unlikely that any individual national regulator could do much more by working alone.

Industry tends to take a more interest-based approach. Most firms with biotechnologies or products of biotechnology assert that the traditional scientific risk assessment approach should be maintained at all costs. Beyond that, they say that product differentiation is their job, and not the job of governments. To that end, they strongly advocate voluntary labelling for GM and other credence factors and either firm-based or sectoral management of any attendant supply chains. As discussed in Chapter 9, the difficulty is that trust is lacking and can only be re-instilled by joint action by industry, government and various collective action groups.

The academic community is not silent in this debate. The advent of biotechnology has created a whole new business in the 'ethics, law, society and industry' (ELSI) of biotechnology. Each discipline has its favourite solution. Sociologists tend to suggest that many of the problems and uncertainties around transformative technologies can be moderated through more, new and better communications and debate. Political scientists tend to use biotechnology to illustrate how public regulatory schemes can come asunder when they ignore the core principles of effective public policy—transparency, accountability and responsibility. Economists, in contrast, offer advice on how firms and industry, in conjunction with governments and associations, can develop new, effective supply chains to differentiate products for different markets. The problem is that each provides too narrow a perspective to resolve the problems of transformative technological change.

Ultimately, each of these approaches has some application, but none alone will resolve the challenge of managing risks and liabilities of a transformative technology. Each would manage one or more risk and liability of the technology, but often at the expense of creating a new risk or liability.

One option is always to do nothing and let the existing systems adapt and resolve any disputes that might arise from the introduction and use of a transformative technology. Without any further action, disputes about the technology would ultimately land in various administrative tribunals and civil courts. While this approach is often the most appropriate for an incremental technology, it is unlikely to be optimal for a transformative technology such as biotechnology. The potential impacts of the technology simply span too many domains. The claims and counterclaims will inevitably conflict. There is a well-known legal dictum that hard cases make bad law. Adjudicating claims related to transformative technologies present perhaps the ultimate hard case. We believe that new institutions (e.g. new norms as discussed in North, 1991) will need to be developed to ensure successful management of the use of this technology.

An Alternative Approach

The underlying theme of this book is that there is no quick fix. While each of the solutions offered or espoused by various interest groups may be appropriate under certain conditions, none by itself will solve the underlying uncertainties created by a transformative technology, nor will they do more than paper over the concomitant socio-economic liabilities.

Complex problems require sophisticated solutions. We believe the management of liabilities resulting from a transformative technology needs to involve a multi-stakeholder strategy. There are inevitably going to be important roles for all of those currently engaged in the debate about the future of agricultural biotechnology and GM foods. The challenge is not to mobilise the various groups, but to engage them in a constructive way.

As suggested in our discussion of institutions in Chapter 1, and throughout this book, we posit that there are three core types of institutions—public, private and collective—and an array of hybrid institutions that mediate and manage aspects of the technology that combine two or more of the core aspects. Each has a role to play.

The public sector has perhaps the most important role. It alone has the capacity to deliver public goods, where there is low rivalry and low excludability. While there is a tendency for governments to do things because they can, it is important that they provide those things that only they can do. In the context of transformative technologies, such as biotechnology, this must involve the establishment of the rules underpinning the system. This most certainly involves creating transparent and accountable structures that deliver credible, predictable, appropriate decisions about technologies and products so that public health and environmental safety are protected. An important adjunct to such a system is open and accessible legal systems to adjudicate disputes in a fair and impartial way. The greatest threat to the public good occurs when private or collective interests capture opaque political or administrative systems and exploit their position to benefit themselves or members of their group. This threat can be lessened by supporting and nurturing well grounded governance in regions, in nation states and through international treaties and organisations. The public governance system will be only as strong as its weakest component.

A vibrant and competitive private sector is equally important in a globalising world. Strong, profitable, accountable firms will be the cornerstone to safe and beneficial development, adaptation, adoption and use of new rival, excludable technologies and products. Firms should be provided with the appropriate legal, commercial and market incentives to act responsibly. The competitive nature of properly structured markets will control most of the feared excesses of private initiative. The recent corporate excesses by a number of global firms have arisen, at least in part, because the rules underpinning corporations have been lax. Management appears often disconnected from owners, and in some firms has ended up using loopholes in accounting, management or governance rules to enrich themselves, to the detriment of their owners' interest. This is not only a failure of markets or private initiative—it is also a result of the failure of the public sector (to provide clear laws prescribing duty of care and duty of performance) and of the collective sectors (e.g. accountants, actuaries and industry regulators in the key bourses to develop and enforce standards of performance). Well structured, profitable, responsible firms are far more likely to deliver the socially optimal amount of technological change than unstable private firms, which are likely either to undershoot or overshoot that goal.

Collective organisations are increasingly important actors. Only they can articulate the views of groups in society—both for-profit groups that need to self-regulate and manage market functions (such as pre-commercial, non-competitive research, standards setting and market development) and not-for-profit social action groups that represent segments of society that cannot be adequately addressed by private or public actors. Voice is important and must be given an opportunity to be mobilised and heard. There is relatively little appreciation for the vital role voice plays in the development and commercialisation of new technologies. Both the public and private sectors need to become more open to engaging with appropriately structured collective actors. Part of that will require new norms to evolve. One problem is that there has been an explosion of collective action—recent studies estimate that there may be more than 60,000 collective organisations around the world seeking to be heard. Not every NGO is voicing the views of a segment of society. Many are simply vehicles for special interests or for profit-making by professional agitators. The public and private sectors need to sort out in their own minds what in their view constitutes a duly-constituted collective actor, and then work with the associations sector to encourage them to develop appropriate governance models. If these organisations want to have their voice heard, then they need to find some way to ensure that they are truly reflecting the views of their members or adherents. Ultimately, these organisations need to be become more transparent and accountable.

Most of the issues resulting from the advent of a transformative technology do not parse cleanly or fully into one of the three pure domains. Rather, they fall in between. The problem is that we have not found good models for dealing with those areas. This book has examined a number of these 'in-between' areas. Chapter 9, for example, discussed the separate and combined roles of the public, private and collective actors in managing product differentiation while Chapter 10 used the advent of plant-made pharmaceuticals to illustrate how challenging those issues might be and how far we are from having an approach to handling them. Clearly delineated roles and responsibilities and appropriate rules and incentives would appear to be critical to handling these in-between issues. Ultimately, though, we need new models of engagement to ensure these issues are dealt with effectively.

In the final analysis, the challenge of managing transformative technological change is not one of developing new science and technology but a problem of organising human relationships. Ultimately, the challenge is less one of material constraints and more one of will. Where there is a will, a way can be found.

Bibliography

Bibliography

Aaker, D. A. (1991) *Managing Brand Equity*. New York: The Free Press.

Agriculture Biotechnology in Europe. (2002) Public attitudes to agricultural biotechnology. Retrieved from the World Wide Web at: www.ABEurope.info.

Agriculture and Environment Biotechnology Commission. (2003) *GM Crops? Coexistence and Liability*. Retrieved from the World Wide Web at: http://www.aebc.gov.uk

Akerlof, G. (1970) The market for lemons: Quality, uncertainty and the market mechanism. *Quarterly Journal of Economics* 84, 488-500.

Allen, J. (1999) Interview with President, Value Added Seeds Ltd., December 17th.

Alston, J., Chan-Kang, C., Marra, M., Pardy, P., and Wyatt, T. (2000) *A Meta-Analysis of the Rates of Returns to Agricultural R&D*. Research Report 113. Washington: International Food Policy Research Institute, September.

Altman, I. and P. Phillips. (2001) Assuring quality in the commodity grain and oilseed trade: Standards and procedures in the Canadian grains regulatory system. Proceedings of the ICABR Meetings, Ravello, Italy, June.

Animal and Plant Health Inspection Services. (2003) Field test releases in the U.S. Retrieved from the World Wide Web at: http://www.nbiap.vt.edu/cfdocs/fieldtests1.cfm.

Anns v. Merton London Borough Council. (1978) A.C. 728 (H.L.) [Anns].

Atkinson, M. (1993) *Governing Canada: Institutions and Public Policy*. Toronto: Harcourt Brace Jovanovich Canada Inc.

Beard, T. (2001) Saskatchewan Wheat Pool. Personal communication. February 5th.

Bender, K. and Hill, L. (2000) *Producer Alternatives in Growing Specialty Corn and Soybeans*. AE-4732. Personally requested. January.

Bender, K., Hill, L., Wenzel, B. and Hornbaker, R. (1999) *Alternative market channels for specialty corn and soybeans*. Retrieved from World Wide Web at: www. ngfa. org/specialtybk. html.

Bhalla, P. L., Swoboda, I., and Singh, M. B. (1999) Antisense-mediated silencing of a gene encoding a major ryegrass pollen allergen. *Proceedings of National Academy of Sciences (USA)* 96, 11676-11680.

Black, H. (1979) *Black's Law Dictionary*, 5th Ed. St Paul: West Publishing Company.

Boyd, S. L., Kerr, W. A. and Perdikis, N. (2003) Agricultural biotechnology innovations versus intellectual property rights – are developing countries at the mercy of multinationals? *The Journal of World Intellectual Property* 6, 2, 211-232.

Brookes, G. (2002) *The farm level impact of using Bt maize in Spain*. Retrieved from the World Wide Web at: http://www.bioportfolio.com/pgeconomics/spain_maize.htm.

Buckingham, D. and Phillips, P.W.B. (2001) Hot potato, hot potato: Regulating products of biotechnology by the international community. *Journal of World Trade* 35, 1, 1-31.

Buckwell, A., Brookes, G. and Bradley, D. (1999). *Economics of identity preservation for genetically modified crops*. Report prepared for Food Biotechnology Communications Initiative.

Bullock, D. S. and Desquilbet, M. (2001) Who pays the costs of non-GMO segregation and identity preservation? Proceedings of the 5th ICABR Meetings, Ravello, Italy, June.

Bullock, D. S., Desquilbet, M. and Nitsi, E. (2000) The economics of non-GMO segregation and identity preservation. Paper presented to the American Agricultural Economics Association Annual Meeting, Tampa, Florida, July 30-August 2.

Button, R. (1999) Communications with Executive Director of Saskatchewan Canola Growers Association, November-December.

Canadian Food Inspection Agency. (2003) Summary of confined research field trials. Retrieved from the World Wide Web at: http://www.inspection.gc.ca/english/plaveg/pbo/triesse.shtml.

Canadian Grain Commission. (1998) *Identity Preserved Systems in the Canadian Grain Industry.* Discussion Paper, December.

Canadian Grains Council. (2002) Food Safety/IP/Traceability Workshop. Winnipeg, Manitoba. February 11.

Canadian Seed Growers Association. (2001) Retrieved from the World Wide Web at: www.seedgrowers.ca.

Canola Council of Canada. (2001) *An Agronomic and Economic Assessment of Transgenic Canola.* Winnipeg. January.

Caulfield, T. (2002) Science, with a Bang: The hype unleashed in the rush to cash in on the genetic revolution threatens to doom important research. Ottawa Citizen A15.

Chèvre, A., Frédérique, E., Jenczweski, E., Darmency, H., and Renard, M. (2000) Gene flow from rapeseed. *Proceedings of the 6th International Symposium on The Biosafety of Genetically Modified Organisms,* Fairbairn, C., Scoles, G. and McHughen, A. (eds) July. Saskatoon, Canada, pp. 45-50. Saskatoon: University Extension Press.

Codex Alimentarius Commission. (2001) Matters arising from Codex committees and task forces: Traceability. Retrieved from the World Wide Web at: ftp://ftp.fao.org/codex/ccexec49/al0121ee.pdf.

Codex Alimentarius Commission. (2003) Joint FAO/WHO Food Standards Programme Codex Alimentarius Commission. Twenty-sixth Session. Rome. Retrieved from the World Wide Web at: http://www.codexalimentarius.net/session_26.stm

Cooke, R. (2001) Grain Marketing Manager, Agricore. Personal communication, January 30[th.]

Comisión Nacional Asesora de Biotecnología Agropecuaria. (2003) Genetically engineered crops: Releases in Argentina. Retrieved from the World Wide Web at: http://siiap.sagyp.mecon.ar/http-hsi/english/conabia/liuk.HTM.

Council of Canadians. (2000) Results of Environics poll on Canadian consumer attitudes to genetically engineered foods. Retrieved from the World Wide Web at: www.canadians.org.

Craven, B., and Johnson, C. (1999) Politics, policy, poisoning and food scares. In: Morris, J., and Bate, R. (eds), *Fearing Food: Risk, Health & Environment.* Oxford, England: Butterworth-Heinemann.

Cruise v. Niessen. (1977) 82 D.L.R. (3d) 190.

Daniell, H. (1999) GM crops: Public perceptions and scientific solutions. *Trends in Plant Science* 4, 12, 467-469.

Dickinson, D. L. and Bailey, D. (2001) Meat traceability: Are U.S. consumers willing to pay for it? *UAES Journal* Paper 7458. Personally requested.

Doern, G. B. and Reed, T. (eds), (2000) *Risky Business: Canada's Changing Science-Based Policy and Regulatory Regime* Toronto: University of Toronto Press, pp. 75-101.

Donaghue v. Stevenson. (1932) A.C. 562 (H.L.).

Eggertsson, T. (1995) Economic perspectives on property rights and the economics of institutions. In Foss, P. (ed), *Economic Approaches to Organizations and Institutions: An Introduction.* Dartmouth, UK: Dartmouth Publishing Co.

Einsiedel, E., Finlay, K. and Arko, J. (2000) *Meeting the public's need for information on biotechnology.* Prepared for Canadian Biotechnology Advisory Committee, October. Retrieved from the WorldWideWeb at: http://www.cbac-cccb.ca/documents/en/PublicNeed_Info_Einsiedel.pdf.

Etzkowitz, H. and Leydesdorff, L. (1995) The triple helix: University-industry-government relations, a laboratory for knowledge based economic development. *European Society for the Study of Science and Technology Review,* 14, 14-19.

EuropaBio. (1997) EuropaBio letter to the President of the Commission of the European Union. Mimeograph.

European Commission. (2003) Deliberate releases and placing on the EU market of Genetically Modified Organisms (GMOs). Retrieved from the World Wide Web at: http://gmoinfo.jrc.it/.

European Union. 2001. Regulation of the European Parliament and of the Council concerning traceability and labelling of genetically modified organisms and traceability of food and feed products produced from genetically modified organisms and amending Directive 2001/18/EC. Retrieved from the World Wide Web at: http://europa.eu.int/comm/food/fs/biotech/biotech09_en.pdf.

European Union Commission. (2000) *Communication from the Commission on the Precautionary Principle* Brussels, 02.02.2000.

Evans, C. (1999) Interview with Executive Director, Biotechnology, Monsanto Canada, November 2.

Ewens, S. and Pusztai, A. (1999) Health risks of genetically modified foods. *Lancet* 354, 684.

Fearne, A. and Garcia, M. (1999) The assured combinable crop scheme in England and Wales: Carrot or Stick? *Farm Management,* 10, 5, 243-261.

Fleming, J. (1983) *The Law of Torts.* 6th Ed. Sydney: The Law Book Company Limited.

Fukuyama, F. (1995) *Trust: The Social Virtues and the Creation of Prosperity.* London: Penguin.

Fulton, M. and Giannakas, K. (2001) Market Effects of Genetically Modified Food. Mimeograph, April.

Gaisford, J. D., Hobbs, J. E., Kerr, W. A., Perdikis, N. and Plunkett, M.D. (2001) *The Economics of Biotechnology.* Cheltenham: Edward Elgar Press.

Gaisford, J. D. and Kerr, W. A. (2001) *Economic Analysis for International Trade Negotiations.* Cheltenham: Edward Elgar Press.

Gaskell, G., Allum, N. and Stares, S. (2003) *Europeans and Biotechnology in 2002.* A Report to the EC Directorate General for Research from the project 'Life Sciences in European Society'. Retrieved from the World Wide Web at http://www.europabio.org.

Gianessi, L., Silvers, C., Sankula, S. and Carpenter, J. (2002) *Plant Biotechnology: Current and potential impact for improving pest management in U.S. agriculture.* Retrieved from the World Wide Web at: www.ncfap.org.

Giddens, A. (1994) Living in a post-traditional society. In Beck, U., Giddens, A., and Lash, S. (eds) *Reflexive Modernity.* Cambridge: Polity.

Ginder, R., Artz, G and Jarboe, D. (2000) *Output trait specialty corn production in Iowa.* Available on the World Wide Web at: www.extension.iastate.edu/Pages/grain/publications/buspub/00gind03.pdf.

Golder, G. and Leung, F. (2000) Economic impact study: Potential costs of mandatory labelling of food products derived from biotechnology in Canada. KPMG Consulting, Ottawa. Available at http://weeds.montana.edu/news/KPMGlabelCanada.pdf.

Good, D., Bender, K. and Hill, L. (2000) *Marketing of Specialty Corn and Soybean Crops.* AE-4733. Personally requested. March.

Gosnell, D. (2001) *Non-GM Wheat Segregation Strategies: Comparing the Costs.* Masters of Science Thesis submitted to the Department of Agricultural Economics, University of Saskatchewan.

Gray, R., Malla, S. and Phillips, P. W. B. (1999) *The Effectiveness of the Research Funding in the Canola Industry.* Regina: Saskatchewan Agriculture and Food.

Gray, R., Malla, S. and Phillips, P. W. B. (Forthcoming) Creating innovative supply chains: A case study of the Canadian canola sector. *Journal of Supply Chain Management.*

Green, J. C. (1971) The Kuhnian Paradigm and the Darwinian revolution in natural history. In: Roller, D. H. D. (ed), *Perspectives in the History of Science and Technology.* Norman, OH: University of Oklahoma Press, 3-25.

Grimsrud, K., McCluskey, J., Loureiro, M. and Wahl, T. (2002) Consumer attitudes toward genetically modified food in Norway. Paper presented to the 6th International Conference of the International Consortium on Agricultural Biotechnology Research (ICABR), Ravello, Italy, July 11-14, 2002.

Gulden, R. H., Shirtliffe, S. J. and Thomas, A. G. (2003) Harvest losses of canola (*Brassica napus*) cause large seedbank inputs. *Weed Science*, 51, 83-86.

Hallman, W. K., Hebden, W. C., Aquino, H. L., Cuite, C. L. and Lang, J. T. (2003) *Public Perceptions of Genetically Modified Foods: A National Study of American Knowledge and Opinion.* (Publication number RR-1003-004). New Brunswick, New Jersey; Food Policy Institute, Cook College, Rutgers – The State of New Jersey.

Harl, N. E., Ginder, R. G., Hurburgh, C. R. and Moline, S. (2000) *The Starlink™ Situation.* Retrieved from the World Wide Web at: http://www.extension.iastate.edu/Pages/grain/publications/buspub/0010star.PDF.

Heller, J. (1995) Quoted in Goudey, J. & D. Nath. 1997. *Canadian Biotech '97: Coming of Age.* Toronto: Ernst & Young.

Hepworth, L. (2001) Industry stewardship as a response to food safety concerns. In Phillips, P. W. B. and Wolfe, R. (eds), *Governing Food: Science, Safety and Trade*, Montreal: McGill University Press/Queens School Studies, 63-74.

Herrman, T., Boland, M. and Heishman, A. (1999) *Economic feasibility of wheat segregation at country elevators.* Retrieved from World Wide Web at: www. css. orst. edu/nawg/1999/herrman. html.

Hobbs, J. E and Kerr, W. A. (1999) Transaction costs. In Bhagwan Dahiya, S. (ed), *The Current State of Economic Science*, 4, Spellbound Publications PVT Ltd, Rohtak, 2111-2133.

Hucl, P. and Matus-Cádiz, M. (2001) Isolation distances for minimizing out-crossing in spring wheat. *Crop Science* 4, 4, 1348-1351.

Hucl, P. (1996) Outcrossing rates in selected Canadian spring wheats. *Canadian Journal of Plant Science* 76, 423-427.

Hudson, L. C., Chamberlain, D. and Stewart, C. N., Jr. (2001) GFP-tagged pollen to monitor pollen flow of transgenic plants. *Molecular Ecology Notes* 1, 321-324.

Huffman, W., Rousu, M., Shogren, J. and Tegene, A. (2002) Should the United States initiate a mandatory labeling policy for genetically modified foods? Paper presented to the 6th International Conference of the International Consortium on Agricultural Biotechnology Research (ICABR), Ravello, Italy, July 11-14.

Hunter v. Canary Wharf Ltd. (1997) NLOR No. 324 NLC 197044701.

Industry Canada. (2000) Strategis: Imports and exports by product. Retrieved from the World Wide Web at: http://strategis.ic.gc.ca/sc_mrkti/tdst/engdoc/tr_homep.html.

Intergovernmental Committee for the Cartegena Protocol on Biodiversity (ICCP) (2001) Liability and Redress for Damage Resulting from the Transboundary Movement of Living Modified Organisms – Note by the Executive Secretary, UNEP/CBD/IPPC/2/3, July 31.

Intergovernmental Committee for the Cartegena Protocol on Biodiversity (ICCP). (2002) Identification of Issues Relating to Liability and Redress for Damage Resulting from the Transboundary Movement of Living Modified Organisms – Note by the Executive Secretary, UNEP/CBD/BS/WS-L&R/1/2, December 2-4.

International Food Information Council. (2003) Americans' Acceptance of Food Biotechnology matches Growers' Increased Adoption of Biotech Crops. Retrieved from the World Wide Web at: HYPERLINK http://www.ific.org/

Isaac, G. E. (2002) *Agricultural Biotechnology and Transatlantic Trade: Regulatory Barriers to GM Crops.* Wallingford, UK: CAB International.

Isaac, G. E. and Kerr, W. A. (2003) GMOs at the WTO: A harvest of trouble. *Journal of World Trade* 37, 6, 1083-1095.

Isaac, G. E., Phillipson, M. and Kerr, W.A. (2002) *International Regulation of Trade in the Products of Biotechnology.* Estey Centre Research Papers No. 2, Estey Centre for Law and Economics in International Trade, Saskatoon.

Jack, D., Pardoe, T. and Ritchie, C. (1998) Scottish quality cereals and coastal grains: Combinable crop assurance in action. *Supply Chain Management* 3, 134-138.

Jacquemin, A. (1987) *The New Industrial Organization.* Cambridge, Mass: The MIT Press.

James, C. (2002) Global Status of Commercialized Transgenic Crops: 2002. *ISAAA Briefs* No. 27. Retrieved from the World Wide Web at: www.isaaa.org.

James, C. and Krattiger, A. F. (1996) Global Review of the Field Testing and Commercialization of Transgenic Plants, 1986 to 1995: The First Decade of Crop Biotechnology. *ISAAA Briefs* No. 1. ISAAA: Ithaca, NY.

Japan Ministry of Agriculture, Forestry and Fisheries. (2003) The Current Status of Transgenic Crop Plants in Japan. Retrieved from the World Wide Web at: http://www.s.affrc.go.jp/docs/sentan/guide/edevelp.htm.

Johnson, C. (2003) IPSOS-REID presentation to the Seed Sector Advisory Committee. December 4.

Kalaitzandonakes, N. (ed) (2003) *Economic and Environmental Impacts of First Generation Biotechnologies.* New York: Kluwer Academic.

Kaye-Blake, W., Bicknell, K. and Lamb, C. (2002) Willingness to pay for genetically modified food labeling in New Zealand. Paper presented to the 6th International Conference of the International Consortium on Agricultural Biotechnology Research (ICABR), Ravello, Italy, July 11-14.

Keller, K. L., (1993) Conceptualising, measuring, and managing customer-based brand equity. *Journal of Marketing* 57 (Jan.), 1-22.

Kennedy, B. (1999) Interview with Business Director of Oilseeds for North America, AgrEvo, October.

Kennedy, K. (1999) Email response to questions from Business Manager, Dow AgroSciences, December 8th.

Kennett, J. (1997) An Examination of Bread Wheat Quality and its Effect on Vertical Co-ordination in the Wheat Supply Chain. Masters of Science Thesis submitted to the Department of Agricultural Economics, University of Saskatchewan.

Kennett, J., Fulton, M., Molder, P. and Brooks, H. (1998) The Case of a UK Baker Preserving the Identity of Canadian Milling Wheat. *Supply Chain Management* 3, 3, 157-166.

Kerr, W. A. (1999) International trade in transgenic food products: A new focus for agricultural trade disputes. *The World Economy* 22, 2, 245-259.

Kerr, W. A. (2002) The international trade regime for biotechnology – a costly muddle. *Business Briefing: Life Sciences Technology* January, 26-29.

Kerr, W. A. (2003a) Chaos or change: Rural participation in the global economy. In: Dalton, G., Bryden, J., Shucksmith, M. and Thomson, K. (eds) *European Rural Policy at the Crossroads*, The Arkleton Centre for Rural Development Research, University of Aberdeen, Aberdeen, pp. 9-23 (also available on line at http://www.abdn.ac.uk/arkleton/conf2000/papers.htm).

Kerr, W. A. (2003b) Science-based rules of trade – a mantra for some, an anathema for others. *The Estey Centre Journal of International Law and Trade Policy* 4, 2, 86-97.

Kerr, W. A and Hall, S.L. (2004) Multilateral environment agreements and agriculture: commitments, cooperation and conflicts. *Current Agriculture, Food and Resource Issues* 5, 39-52.

Kerr, W. A. and Hobbs, J. E. (2002) The North American-European Union dispute over beef produced using growth hormones: A major test for the new international trade regime. *The World Economy* 25, 2, 283-296.

Khachatourians, G. G. (2002) Agriculture and Food Crops: Development, science and society. In: Khachatourians, G. G., McHughen, A., Nip, W-K., Scorza, R. and Hui, Y-H. (eds), *Transgenic Plants and Crops*. New York: Marcel Dekker Inc, 1-27.

Khachatourians, G. G. (2001) How well understood is the "science" of food safety. In: Phillips, P. W. B. and Wolfe, R. (eds), *Governing Food: Science, Safety and Trade*. Montreal: McGill-Queen's University Press, 13-26.

Khachatourians, G. G., McHughen, A., Nip, W-K., Scorza, R. and Hui, Y-H. (2002) (eds), *Transgenic Plants and Crops*. New York: Marcel Dekker Inc.

Kuhn, T. (1970) *The Structure of Scientific Revolutions*. 2nd Edition, Chicago: University of Chicago Press.

Kuntz, G. (2001) Transgenic Wheat: Potential Price Impacts for Canada's Wheat Export Market. Masters of Science Thesis submitted to the Department of Agricultural Economics, University of Saskatchewan.

Kwaczek, A. S., Kerr. W. A. and Mooney, S. (1990) Chernobyl: Lessons in nuclear liability. *Forum for Applied Research and Public Policy* 5, 2, 21-27.

Lane, V. and Jacobson, R. (1995) Stock market reactions to brand extension announcements: The effects of brand attitude and familiarity. *Journal of Marketing* 59 (Jan.) 63-77.

Lefol, E., Danielou, V. and Darmency, H. (1996a) Predicting hybridization between transgenic oilseed rape and wild mustard. *Field Crops Research* 45, 153-161.

Lefol, E., Fleury, A. and Darmency, H. (1996b) Gene dispersion from transgenic crops II. Hybridization between oilseed rape and the wild hoary mustard. *Sexual Plant Reproduction* 9, 186-196.

Lehnert, S., Schmitz, T. and Peterson, B. (2000) Risk and weak point analysis in the range of chain-oriented data acquisition. In: Trienekens, J. H. and Zuubier, P. J. P. (eds) *Chain Management in Agribusiness and the Food Industry*. Wageningen, The Netherlands: Wageningen Pers, 409-416.

Leiss, W. (2001) *Understanding Risk Controversies*. Montreal: McGill-Queen's University Press.

Lin, W. W. (2002) Estimating the costs of segregation for non-biotech maize and soybeans. In: Santaniello, V., Evenson, R. E. and Zilberman, D. (eds), *Market Development for Genetically Modified Foods*. Wallingford, UK: CAB International.

Lin, W. W. and Johnson, D. D. (2004) Segregation of non-biotech corn and soybeans: Who bears the cost? In: Santaniello, V. and Evenson, R. E. (eds), *The Regulation of Agricultural Biotechnology*. Wallingford, UK: CAB International.

Lin, W. W., Price, G. K. and Allen, E. W. (2003) StarLink: Impacts on the U.S.Corn market and World Trade. *Agribusiness* 19, 4, 473-488.

Losey, J. (1999) Transgenic pollen harms monarch larvae. *Nature* 399, 6733.

Ma, J. K-C., Drake, P. M. W. and Christou, P. (2003) The production of recombinant pharmaceutical proteins in plants. *Nature Genetics* 4, 794-805.

Maltsbarger, R. and Kalaitzandonakes, N. (2000) Direct and hidden costs of identity preservation. *AgBioForum* 3, 4, 236-242.

Manitoba Pool Elevators. (1996) Costs of Identity Preservation. (Mimeograph).

Mayer, H. L. (1997) The Economics of Transgenic Herbicide-Tolerant Canola. Masters of Science Thesis submitted to the Department of Agricultural Economics, University of Saskatchewan.

Mayer, H. L. and Furtan, H. (1997) Economics of transgenic herbicide-tolerant canola: The case of Western Canada. *Food Policy* 24, 431-442.

McCloskey, R. (2002) *Presentation to the Pew Initiative on Food and Biotechnology Conference*. Retrieved from the World Wide Web at: http://pewagbiotech.org/events/0717/ConferenceReport.pdf.

McCluskey, J., Ouchi, H., Grimsrud, K. and Wahl, T. (2001) *Consumer response to genetically modified food products in Japan*. Retrieved from the World Wide Web at: www.impact.wsu.edu. TWP-2001-101.

McHughen, A. (2000) *Pandora's Picnic Basket: The Potential and Harzards of Genetically Modified Foods*. Oxford: Oxford University Press.

Mehta, M. D. (2003) Regulating biotechnology and nanotechnology in Canada: A post-normal science approach for inclusion of the fourth helix. In Baber, Z. and Khondker, H. (eds), *The Triple Helix*. Albany, N.Y.: State University of New York Press.

Mihalchuk v. Ratke. (1966) 57 D.L.R. (2d) 269.

Mills, L. (2002) *Science and Social Context: The Regulation of Recombinant Bovine Growth Hormone in North America*. Montreal: McGill-Queen's University Press.

Miranowski, J. A., Moschini, G., Babcock, B., Duffy, M., Wisner, R., Beghin, J., Hayes, D., Lence, S., Baumel, C. P. and Harl, N. E. (1999) *Economic perspectives on GMO market segregation*. Staff Paper 298. Iowa State University. Retrieved from the World Wide Web at: http://agecon.lib.umn.edu/cgi-bin/detailview.pl?paperid=1768.

Moon, W. and Balasubramanian, S. (2002) Public perceptions and willingness-to-pay a premium for non-GM foods in the US and UK. *AgBioForum* 4, 3, 221-231.

MORI (2002) British Public Supports EU Directive. Retrieved from the World Wide Web at: www.mori.com/polls/2002/greenpeace.shtml.

Moschini, G., Lapan, H. and Sobolevsky, A. (1999) Roundup Ready Soybeans and Welfare Effects in the Soybean Complex. *Agribusiness* 16, 33-55.

Moss, C. B., Schmitz, T. G. and Schmitz, A. (2004) Differentiating GMOs and Non-GMOs in a Marketing Channel. In: Santaniello, V. and Evenson, R. E. (eds), *The Regulation of Agricultural Biotechnology*. Wallingford, UK: CAB International.

Murphy, J. M. (1990) Assessing the value of brands. *Long Range Planning* 23, 3, 23-29.

National Academy of Sciences. (1983) *Risk Assessment in the Federal Government: Managing the Process*. Committee on the Institutional Means for Assessment of Risks to Public Health, Commission of Life Sciences, National Academy Press, Washington.

National Research Council. (1998) Commercialisation of plants with novel traits (PBI Bulletin, September). Retrieved from the World Wide Web at: http://www.pbi.nrc.ca/bulletin/sept98/production.html.

Natural Sciences and Engineering Research Council. (2001) Synergy Awards for R&D Partnerships. Retrieved from the World Wide Web at: www.nserc.ca/program/synerg/2001_breed_e.htm.

New Zealand. (2001) *Report of the Royal Commission on Genetic Modification*. Wellington: Printlink.

North, D. C. (1991) Institutions. *Journal of Economic Perspectives* 5, 97-112.

North, D. C. (1990) *Institutions, Institutional Change and Economic Performance*. Cambridge: Cambridge University Press.

Odell, J. T. (1991) Site specific recombination in plant cells. Patent WO 91/09957.

Office of the Gene Technology Regulator (2003) Record of GMOs and GM Products – Licences involving an intentional release of GMOs into the environment. Retrieved from the World Wide Web at: http://www.ogtr.gov.au/gmorec/ir.htm.

Olson, M. (1965) *The Logic of Collective Action: Public Goods and the Theory of Groups*. London: Harvard University Press.

Organization of Economic Cooperation and Development. (2000) Regulatory Developments in Biotechnology in OECD Member countries. Retrieved from the World Wide Web at http://www.oecd.org/ehs/country.htm.

Organization of Economic Cooperation and Development. (2003) OECD's Database of Field Trials. Retrieved from the World Wide Web at: http://www.olis.oecd.org/biotrack.nsf.

Osborne, P. (2000) *The Law of Torts*. Retrieved from the World Wide Web at: http://www.quicklaw.com.

Overseas Tankship (U.K.) Ltd. v. Morts Dock and Engineering Co. Ltd. The Wagon Mound (No.1) (1961) A.C. 388 (P.C.).

Perdikis, N and Kerr, W. A. (1999) Can consumer-based demands for protection be incorporated in the WTO? – The case of genetically modified foods. *Canadian Journal of Agricultural Economics* 47, 4, 457-465.

Pew Initiative on Food and Biotechnology. (2002) *Pharming the Field: A look at the benefits and risks of bioengineering plants to produce pharmaceuticals*. Retrieved from the World Wide Web at: http://pewagbiotech.org/events/0717/ConferenceReport.pdf.

Philips v. California Standard Co. (1960) 31 W.W.R. 331 (Alta. S.C.).

Phillips, P. W. B. and McNeill, H. (2002) Labeling for GM foods: Theory and Practice. In: Santaniello, V., Evenson, R. and Zilberman, D. (eds), *Market Development for Genetically Modified Foods*. Wallingford, UK: CAB International.

Phillips, P. W. B. and Kerr, W. A. (2000) Alternative paradigms – the WTO versus the Biosafety Protocol for trade in genetically modified organisms. *Journal of World Trade* 34, 4, 63-75.

Phillips, P. W. B. and Kerr, W. A. (2002) Frustrating competition through regulatory uncertainty – international trade in the products of biotechnology. *World Competition Law and Economics Review* 25, 1, 81-99.

Phillips, P. W. B. and Smyth, S. (Forthcoming) Managing the value of new-trait varieties in the canola supply chain in Canada. *Supply Chain Management.*

Phillips, P. W. B. (2002) Regional systems of innovation as a modern R&D entrepot: The case of the Saskatoon biotechnology cluster. In: Chrisman, J., *et al.* (eds), *Innovation, Entrepreneurship, Family Business and Economic Development: A Western Canadian Perspective.* Calgary: University of Calgary Press, 31-58.

Phillips, P. W. B. and Khachatourians, G. G. (2001) *The Biotechnology Revolution in Global Agriculture: Invention, Innovation and Investment in the Canola Sector.* Wallingford, U.K: CAB International.

Phillips, P. W. B. and Wolfe, R. (2001) Governing food in the 21st century: The globalization of risk analysis. In: Phillips, P. W. B. and Wolfe, R. (eds), *Governing Food: Science, Safety and Trade.* Montreal: McGill University Press/Queens School Studies, 1-10.

Picciotto, R. (1995) *Putting Institutional Economics to Work: From Participation to Governance.* World Bank Discussion Paper 304. Washington, D.C.

Plunkett, M. D. and Gaisford, J. D. (2000) Limiting biotechnology? Information problems and policy responses. *Current Agriculture, Food and Resource Issues* 1, 21-28. Retrieved from the World Wide Web at: http://www.CAFRI.org.

Pohl-Ort, M., Brand, U., Drieben, S., René-Hesse, P., Lenhen, M., Morak, C., Mücher, T., Saeglit, C., von Soosten, C. and Bartsch, D. (2000) Overwintering of genetically modified sugar beet, *Beta vulgaris* L. subsp. *vulgaris*, as a source for dispersal of transgenic pollen. *Euphytica* 108, 181-186.

Price, G. K., Kuchler, F. and Krissoff, B. (Forthcoming) E.U. Traceability and the U.S. Soybean Sector. In: Santaniello, V. and Evenson, R. E. (eds), *The Regulation of Agricultural Biotechnology.* Wallingford, UK: CAB International.

Produce Marketing Association. (2001) *Fresh trends: Understanding consumers and produce.* Retrieved from the World Wide Web at: www.pma.com.

Quist, D. and Chapela, I. (2001) Transgenic DNA introgressed into traditional maize landraces in Oaxaca, Mexico. *Nature* 414, 541-543.

Reichert, H. and Vachal, K. (2000) *Identity preserved grain: Logistical overview.* Retrieved from the World Wide Web at: http://151.121.3.151/tmd/IPGrain/ipgrains.pdf.

Rousu, M., Huffman, W., Shogren, J. and Tegene, A. (2002) Are US consumers tolerant of GM foods? Paper presented to the 6th International Conference of the International Consortium on Agricultural Biotechnology Research (ICABR), Ravello, Italy, July 11-14.

Rylands v Fletcher. (1868) L.R. 3 H.L. 330, affg L.R. 1 Ex. 265.

Sackett, D. L, Rosenberg, W. M. C., Gray, J. Å. M., Haynes, R. B. and Richardson, W. S. (1996) Evidence based medicine: What it is and what it isn't. *British Medical Journal* 312, 71-72.

Saeglitz, C., Pohl, M. and Bartsch, D. (2000) Monitoring gene flow from transgenic sugar beet using cytoplasmic male-sterile bait plants. *Molecular Ecology* 9, 2035-2040.

Sandman, P. M. (1994) Mass media and environmental risk: Seven principles. In: *Risk: Health, Safety, and Environment*, Summer, 1-7.

Santaniello, V., Evenson, R., Zilberman, D. and Carlson, G. (2000) *Agriculture and Intellectual Property Rights: Economic, institutional and implementation issues in biotechnology.* Wallingford, UK: CAB International.

Saskatchewan Wheat Pool (1997) Identity preserved canola handling in Canada. (Mimeograph).

Sawhney, V. K. (2001) Pollen Biotechnology. In: Khachatourians, G., G., McHughen, A., Nip, W-K., Scorza, R.and Hui, Y-H.(eds), *Transgenic Plants and Crops*. New York: Marcel Dekker Inc.

Shrader-Frechette, K. (1990) *Perceived risks versus actual risks: Managing hazards through negotiations*. Retrieved from the World Wide Web at: http://www.piercelaw.edu/Risk/Vol1/fall/ShraderF.htm.

Shapiro, R. (1999) Open letter from Monsanto CEO Robert B. Shapiro to Rockefeller Foundation President Gordon Conway. October 4th. Retrieved from the World Wide Web at: www.monsanto.com/monsanto/gurt/default.htm.

Slusar, T. (2002) Personal communication with Grain Manager for CanAmera Foods.

Smyth, S. and Phillips, P. W. B. (2001) *Identity-preserving production and marketing systems in the global agri-food market: Implications for Canada*. Regina: Saskatchewan Agriculture and Food.

Smyth, S. and Phillips, P. W. B. (2002) Competitors co-operating: Establishing a supply chain to manage genetically modified canola. *International Food and Agribusiness Management Review* 4, 1, 51-66.

Smyth, S. and Phillips, P. W. B. (2003) Product differentiation alternatives: Identity preservation, segregation and traceability. *AgBioForum*, 5, 2, 30-42.

Smyth, S., Khachatourians, G. G. and Phillips, P. W. B. (2002) Liabilities and Economics of Transgenic Crops. *Nature Biotechnology* 20, 537-541.

Snow, A., Pilson, D., Rieseberg, L., Paulsen, M., Pleskac, N., Reagon, M., Wolf, D. and Selbo, S. (2003) A Bt transgene reduces herbivory and enhances fecundity in wild sunflowers. *Ecological Applications*, 13, 2, 279-286.

Snow, E. (1997) Cited by Craven, B., and Johnson, C., 1999, Politics, policy, poisoning and food scares. In: Morris, J., and Bate, R. (eds), *Fearing Food: Risk, Health & Environment*. Oxford, England: Butterworth-Heinmann.

Sparks Companies, Inc. (2000) *The IP future: Identity Preservation in North American agriculture*. Memphis.

Spriggs, J. and Isaac, G. E. (2001) *International Competitiveness and Food Safety: The Case of Beef*. Wallingford, UK: CAB International.

Stanbury, W. T. (2000) Reforming Risk Regulation in Canada. In: Jones, L. (ed.) *Safe Enough? Managing Risk and Regulation*. Vancouver: The Fraser Institute.

Staniland, B. K., McVetty, P. B. E., Friesen, L. F., Yarrow, S., Freyssinet, G. and Freyssinet, M. (2000) Effectiveness of border areas in confining the spread of transgenic *Brassica napus* pollen. *Canadian Journal of Plant Science* 80, 521-526.

Stiglitz, J. (1999) Wither Reform? Ten Years of the Transition. World Bank Annual Bank Conference on Development Economics, April 28-30.

Strange, S. (1988) *States and Markets*. London: Pinter.

The Agricultural Operations Act. (1995) Regina, Saskatchewan: Queen's Printer.

Timon, D. and O'Reilly, S. (1998) An evaluation of traceability systems along the Irish beef chain. In Viau, C. (ed), *Long-term Prospects for the Beef Industry*. INRA: Ivry-sur-Seine, 219-225.

Tirole, J. (1988) *The Theory of Industrial Organization*. Cambridge, Mass.: MIT Press.

Tomes, D. T. (1997) Genetic constructs and methods for producing fruits with very little or diminished seed. Patent WO97/40179.

Traxler, G., Godoy-Avila, S., Falck-Zepeda, J. and Espinoza-Arellano, J. (2003) Transgenic cotton in Mexico: Economic and environmental impacts. In: Kalaitzandonakes, N. (ed) *Economic and Environmental Impacts of First Generation Biotechnologies*. New York: Kluwer Academic.

van den Belt, H. (2003) Debating the Precautionary Principle: "Guilty until proven innocent" or "innocent until proven guilty"? *Plant Physiology* 132, 1122-1126.

van den Daele, W., Puhler, A., and Sukopp, H. (1997) *Transgenic Herbicide- Resistant Crops: A Participatory Technology Assessment.* Summary Report for the Federal Ministry for Research and Technology. Berlin: Wissenschaftszentrum Berlin fur Sozialforschung.

Vido, E., Ojah, M. and Kosior, J. (2000) *A Looming Crisis for Western Canadian Grain?* Available on the World Wide Web at: www.umanitoba.ca/transport_institute/publications/CTRF%202000%20Paper.pdf.

Visser, B., Eaton, D., Louwaars, N. and van der Meer, I. (2001) *Potential impacts of genetic use restriction technologies (GURTs) on agrobiodiversity and agricultural production systems.* Study carried out for FAO, April.

Washington Post. (2003) Firm fined for spread of altered corn genes. April 24th, p. E4.

Wasylyniuk, C. R., Bessel, K. M., Kerr, W. A. and Hobbs, J. E. (2003) The Evolving International Trade Regime for Food Safety and Environmental Standards: Potential Opportunities and Constraints for Saskatchewan's Beef Feedlot Industry. *Estey Centre for Law and Economics in International Trade.*

Western Producer, The. (2000) Triple-resistant canola weeds found in Alta. Retrieved from the World Wide Web at: www.producer.com/articles/20000210/news/20000210news01.html.

Williamson, O. E. (1979) Transaction-cost economics: The governance of contractual relations. *Journal of Law and Economics* 22, 233-261.

Wilson, N. and Clarke, W. (1998) Food safety and traceability in the agricultural supply chain: Using the Internet to deliver traceability. *Supply Chain Management* 3, 127-133.

Wolf, M. and Kari, B. (2001) A comparison of consumer attitudes toward genetically modified food in the United States over four time periods. Paper presented to the 5th International Conference of the International Consortium on Agricultural Biotechnology Research (ICABR), Ravello, Italy, June 15-18.

Wolf, M. and Pachico, D. (2002) Attitudes toward genetically modified food in Colombia. Paper presented to the 6th International Conference of the International Consortium on Agricultural Biotechnology Research (ICABR), Ravello, Italy, July 11-14.

Wolf, M., Stephens, A. and Pedrazzi, N. (2001) Using simulated test marketing to examine purchase interest in food products that are positioned as GMO free. Paper presented to the 5th International Conference of the International Consortium on Agricultural Biotechnology Research (ICABR), Ravello, Italy, June 15-18.

Wolf, M., Bertolini, P. and Parker-Garcia, J. (2002) A comparison of consumer attitudes toward genetically modified food in Italy and the Unites States. Paper presented to the 6th International Conference of the International Consortium on Agricultural Biotechnology Research (ICABR), Ravello, Italy, July 11-14.

World Trade Organisation. (1998) *Appellate Decision, Australia - Measures Affecting Importation of Salmon - Complaint by Canada.* AB-1998-5. (Appeal from WT/DS18. Available online at http://wto.org.).

Zucker, L., Darby, M. and Brewer, M. (1998) Intellectual human capital and the birth of U.S. biotechnology enterprises. *American Journal of Economics* 88, 1, 290-306.

Index